Fish and Fish Products

Third Supplement to the Fifth Edition of

McCance and Widdowson's

The Composition of Foods

Fish and Fish Products

Third Supplement to the Fifth Edition of

McCance and Widdowson's

The Composition of Foods

B. Holland, J. Brown and D.H. Buss

The Royal Society of Chemistry
and
Ministry of Agriculture, Fisheries and Food

The Royal Society of Chemistry
Thomas Graham House
Science Park
Milton Road
Cambridge CB4 4WF
UK

Tel.:   +44 (0) 223 420066   Telex: 818293
Fax: +44 (0) 223 423429

ISBN 0-85186-421-X

Orders should be addressed to:
The Royal Society of Chemistry
Turpin Distribution Services Ltd
Blackhorse Road
Letchworth
Herts. SG6 1HN

Tel.:   +44 (0) 462 67255   Telex: 825372
Fax: +44 (0) 462 480947

Photocomposed by Land and Unwin (Data Sciences) Limited, Bugbrooke

Printed in the United Kingdom at Clays Ltd St Ives plc

CONTENTS

ACKNOWLEDGEMENTS

A large number of people have helped at each stage in the preparation of this book.

Most of the new analyses of fish and fish products were performed at the Laboratory of the Government Chemist, Teddington. The analytical team was headed by Mrs G D Holcombe. A number of samples of fish from fish and chip shops were purchased by members of the Procter Department of Food Science, University of Leeds, under the direction of Mrs J Ryley.

We are indebted to numerous manufacturers, retailers and other organisations for information on the range and composition of their products. In particular we would like to thank Birds Eye Wall's Limited., Bluecrest Foods Ltd, John West, Lyons Seafoods Ltd, Ross Young's Ltd, J Sainsbury plc, Sea Products International, the Seafish Industry Authority, Tesco Stores Ltd, Van den Bergh Professional Foods and Waitrose Ltd. We would especially like to thank Mrs Husneara Hussein and Dr Sandra Warrington from the Royal London Hospital, Whitechapel, dietitians from Gardner Merchant Ltd, members of Torry Research Station and Dr Angus Aitken (formerly Torry Research Station) for their valuable contributions to the contents of this supplement. The Seafish Industry Authority kindly provided the cover photographs.

The final preparation of this book was overseen by a committee which, besides the authors, comprised Dr M C Edwards (Campden Food and Drink Research Association, Chipping Campden), Dr A M Fehily (H J Heinz Company Ltd), Miss A A Paul (MRC Dunn Nutrition Centre, Cambridge), Mrs P M Richardson (Northwick Park Hospital, Harrow), Professor D A T Southgate (formerly AFRC, Institute of Food Research, Norwich), and Dr F J Taylor (Royal Society of Chemistry, Cambridge). Miss P M Brereton retired as a member of the committee during work on this supplement. We are indebted to her for her helpful contributions to this and earlier publications.

We would also like to express our appreciation for all the help given to us by many people in the Ministry of Agriculture, Fisheries and Food, The Royal Society of Chemistry and elsewhere who were involved in the work leading up to the production of this book. In particular, we would like to thank Mrs A A Welch (formerly Royal Society of Chemistry, Cambridge) for her help in the early stages of production of this supplement and Dr W Chan who recently joined the team.

INTRODUCTION

This book presents information on up to 79 different nutrients, including minerals, vitamins and a wide range of fatty acids, in 308 fish and fish products. It is the eighth supplement to McCance and Widdowson's *The Composition of Foods*, and the sixth in a series showing the most recent information on the nutritional composition of specific groups of foods.

The fifth edition of *The Composition of Foods* was published in 1991 by the Royal Society of Chemistry (RSC) and the Ministry of Agriculture, Fisheries and Food (MAFF) (Holland *et al.*,1991b), and shows the amounts of the major nutrients in 1188 of the more commonly eaten foods in Britain. The RSC and MAFF are also producing a series of supplementary reference books on particular groups of foods, with more detailed information on a much wider range of foods within each group. Previous supplements have covered *Cereals and Cereal Products* (Holland *et al.*,1988), *Milk Products and Eggs* (Holland *et al.*,1989), *Vegetables, Herbs and Spices* (Holland *et al.*,1991a), *Fruit and Nuts* (Holland *et al.*,1992a, and *Vegetable Dishes* (Holland *et al.*,1992b). All the information from the supplements, including this one is also available in computer-readable format, details of which can be obtained from the Royal Society of Chemistry. Each of these supplements is an essential reference work for those who need the most up to date information on the widest possible range of foods and nutrients.

The present book continues this tradition, and shows the nutrients in a much wider range of fresh and cooked fish and fish products than the 97 in the fifth edition. It also includes information on a wider range of nutrients, including the individual fatty acids which are of particular importance in certain fish, and the main fractions of vitamin A and vitamin E. All the data have been thoroughly reviewed and almost every value has been updated. This was necessary because most of the data on fish in the fifth edition were taken directly from the third and fourth editions, and had thus not been reassessed or revised for many years.

Sources of data

The selection of foods and of nutrient values has followed the general principles used in the preparation of previous supplements and the fifth edition of *The Composition of Foods*. As far as possible, all the major fish and fish products eaten in the UK have been included, and most of the nutrient values are derived from new analyses of representative samples of each food. All the samples were bought from typical retail outlets, and each was analysed especially for this supplement, both raw and after cooking by a range of methods as appropriate. Some of the values were, however, taken from the scientific literature as described below.

Literature values

Before any analytical work was commissioned, an extensive review of the world's scientific literature was undertaken. This gave values for a number of nutrients in many of the fish and fish products eaten in the UK, and selected values have been included where the studies gave full details of the samples, where suitable methods of analysis had been used, and where the results were presented in sufficient detail for a full evaluation to be made. In the absence of other data, however, literature values for a number of fish imported from the Bangladeshi, Caribbean and Mediterranean regions were included directly, although preference was given to values where the species of fish were those eaten in the UK and where appropriate cooking methods had been used.

The review showed that many gaps remained, and for these as well as for new products and for the fish and fish products now most commonly eaten in the UK, a large number of new analyses were commissioned.

New analyses

For fish and fish products for which there were no recent or reliable values, or where the earlier values related to varieties that are no longer widely available, arrangements were made for samples to be bought, prepared and cooked if appropriate, and then analysed at the Laboratory of the Government Chemist. Two large studies were undertaken. In each case representative samples were bought from a variety of retail outlets, including supermarkets, fishmongers, fish stalls and department stores, mainly in the London area. In addition, samples of many cooked fish were purchased from fish and chip shops in Leeds as well as in London. Up to 24 different samples of each fish were bought from different outlets and then combined before analysis by the methods indicated below.

For the items that needed to be cooked before analysis, fresh fish were coated if necessary and cooked by typical methods, while processed fish and fish products were cooked according to the instructions on the packet. Where frying was done in more than one oil or fat, the fat used is specified. "Retail blend oil" is a weighted average of the frying oils used in catering and in fish and chip shops.

The analytical methods for the major nutrients were as described in the fifth edition of *The Composition of Foods*, while those for the nutrients not included in that book are given in the supplements on *Milk Products and Eggs* or on *Vegetables, Herbs and Spices*. Individual fatty acids were determined as their methyl esters by capillary gas chromatography. Further details of individual determinations can be provided on request.

Arrangement of the tables

Food groups

For ease of reference, the values for white fish and cartilaginous fish have been brought together, and are then followed by values for fatty fish, crustacea, molluscs, and finally by a section on fish products and fish dishes. Although the foods have been listed alphabetically within each section, the first value for each is for the fresh or raw fish, followed where appropriate by values for the cooked fish and canned products.

The compositions of cartilaginous fish (shark, skate and rock salmon/dogfish) differ from those of other white fish in that they contain substantial amounts of non-protein nitrogen (particularly urea) in the muscle. Nevertheless, they have been brought together in this supplement because they are normally used interchangeably in the diet. There are greater compositional differences between this group and fatty fish, so fatty fish have been kept separate. It should, however, be appreciated that some "fatty fish" have very little fat; that the amount can differ markedly with season; and that much of the fat in some fatty fish is located in dark muscle which is immediately under the skin and may therefore not be eaten.

Numbering system

As in previous supplements, the foods have been numbered in sequence, with white and cartilaginous fish numbered from 1 to 167, fatty fish from 168 to 231, crustacea from 232 to 249, molluscs from 250 to 271, and fish products and dishes from 272 to 308. Each food in the computerised MAFF/RSC nutrient databank also has a unique two digit prefix to specify its food group, and for this supplement the prefix is '16'. The full code numbers for raw *ayr* and *tuna pâté*, the first and last foods in this supplement, are thus 16-001 and 16-308, and these are the numbers that will be used in other databank applications.

Some fish are known by a number of different names. Each is listed in the main table under its most common name, but all the names have been included in the index and there is also an Appendix of alternative names and scientific names on page 123 to help with identification.

Since different code numbers have been used for convenience in the fifth edition, a full list will be available from the Royal Society of Chemistry which shows for all foods the numbers used in this and the other supplements as well as the numbers used in the fifth edition.

Description and number of samples

The information given under this heading indicates the number and nature of the samples taken for analysis. Some additional values for cooked fish were calculated from these, usually after correction for differences in water content and nutrient losses as described in the fifth edition of *The Composition of Foods*. For example, the calculations for grilled fish, were as follows:

$$\text{Composition of cooked fish per 100g} = \frac{\text{Nutrients in weight of raw fish}}{\text{Weight of grilled fish}} \times 100$$

$$\text{Water content of cooked fish per 100g} = \frac{\text{Water in raw fish} - \text{weight loss on cooking}}{\text{Weight of grilled fish}} \times 100$$

The weight losses on grilling were measured and taken to be water since such calculations were needed only for white fish. Such calculations are indicated under the heading, as is the type of oil used if the fish was fried.

Where all the nutrient values for a fish were derived from a single literature source, this source is indicated, but not where literature values were used primarily for confirmation or where selected values were combined with the new analytical data. A full list of the 244 references consulted can be obtained on request.

Nutrients

For each food, the nutrient values have been given per 100 grams of the edible portion of the food. However, fish are often served and weighed with varying amounts of inedible material (small or large bones, heads, tails and skin), so, for a number of the more popular fish, values have also been given per 100 g of the fish as commonly served. The inedible material is described and quantified in each case, but because the amounts can be extremely variable, users of these tables should ensure that their samples reflect those described before applying the values in this supplement.

The presentation of the nutrients differs slightly from that in previous supplements in order to show within four pages the nutritional information most appropriate to fish.

Proximates: - The first page for each food begins with the proportion of edible matter in the food *as described.* When the skin or bones have been included as edible, this is stated. The information continues with the amounts of water, total nitrogen, protein, fat, available carbohydrate expressed as its monosaccharide equivalent, and the energy value of the food both in kilocalories and kilojoules.

The protein content of most fish and fish products was derived by multiplying the total amount of nitrogen by 6.25, but for cartilaginous fish, which contain a substantial amount of non-protein nitrogen, this nitrogen was first deducted. As in all previous UK food tables, the amounts of starch and sugars were determined analytically and not estimated 'by difference'. Some shellfish contained glycogen, and the amounts are included in the total carbohydrate but not in the columns for starch or sugars. Energy values were derived by multiplying the amounts of protein, fat and total carbohydrate by the factors in Table 1.

Table 1   *Energy conversion factors*

	kcal/g	kJ/g
Protein	4	17
Fat	9	37
Available carbohydrate expressed as monosaccharide	3.75	16

Carbohydrates, fibre and fatty acids: – The second page gives the amounts of starch and total sugars, and two measures of fibre (non-starch polysaccharides, determined by the method of Englyst and Cummings (1988), and unavailable carbohydrate as determined by Southgate (1969)). These components are normally present only in fish products and some cooked fish from the other ingredients used. As in previous supplements, the amounts of starch and sugars are shown after conversion to their monosaccharide equivalents, while the amounts of fibre are the actual weights of the material. In addition, the total amounts of saturated, monounsaturated and polyunsaturated fatty acids and cholesterol are shown. More details of the individual fatty acids in the main raw fish are given in the Appendix on page 101. To convert the cholesterol values to millimoles, divide by 386.6.

Minerals and vitamins: - The range of minerals and vitamins shown in the main tables is the same as in the fifth edition of *The Composition of Foods* except for sulfur. The amounts of calcium and phosphorus may vary in fish as consumed from the values determined in the samples analysed for this supplement, because of the difficulty of removing all the fine bones from some fish. The amounts of iron, copper and zinc can also vary widely in those crustacea and molluscs that can accumulate large amounts of these and heavy metals from their environment. Most nutrient values were obtained by direct analysis, but for some cooked fish selected values were calculated from the corresponding raw foods. For these, any vitamin losses were estimated from the losses found on analysis of similar foods, as shown in Table 2. The amounts of retinol and vitamin E given take account of the relative activities of the different fractions where these are known. The amounts of these fractions are shown on page 114–119.

Table 2 *Typical vitamin losses (%) on cooking fish[a]*

	Poaching and steaming	Baking	Grilling	Frying
Vitamin A	0	0	0	0
Vitamin D	0	0	0	0
Vitamin E	0	0	0	0
Thiamin	10	30	10	20
Riboflavin	0	20	10	20
Niacin	10	20	10	20
Vitamin B_6	0	10	10	20
Vitamin B_{12}	0	10	0	0
Folate	0	20	0	0
Pantothenic acid	20	20	5	20
Biotin	10	10	0	10

[a] Apart from grilling, the losses are mainly based on those found on cooking cod

Appendices

There are six appendices. The first, on page 101, shows the individual fatty acids per 100 g of those fish and fish products with more than 0.5 g fat per 100 g, and the second appendix on page 112, shows the amounts of individual sugars in those fish and fish products where these were analysed. Most are derived from other ingredients such as batter. There may also be glucose in shellfish arising from the natural breakdown of glycogen. The third and fourth appendices on pages 114 and 116 show the amounts of the major fractions of vitamin A and vitamin E in those fish that contain them. There is a further appendix of recipes for the fish dishes included in this supplement, followed by a list of the main alternative and taxonomic names for fish (page 123).

References to introductory text

Englyst, H. N., and Cummings, J. H. (1988) An improved method for the measurement of dietary fibre as the non-starch polysaccharides in plant foods. *J. Assoc. Off. Analyt. Chem.* **71**, 808-814

Holland, B., Unwin, I. D., and Buss, D. H. (1988) *Third supplement to McCance and Widdowson's The Composition of Foods: Cereals and Cereal Products*, Royal Society of Chemistry, Cambridge

Holland, B., Unwin, I. D., and Buss, D. H. (1989) *Fourth supplement to McCance and Widdowson's The Composition of Foods: Milk Products and Eggs*, Royal Society of Chemistry, Cambridge

Holland, B., Unwin, I. D., and Buss, D. H. (1991a) *Fifth supplement to McCance and Widdowson's The Composition of Foods: Vegetables, Herbs and Spices*, Royal Society of Chemistry, Cambridge

Holland, B., Welch, A. A., Unwin, I. D., Buss, D. H., Paul, A. A., and Southgate, D. A. T. (1991b) *McCance and Widdowson's The Composition of Foods 5th edition*, Royal Society of Chemistry, Cambridge

Holland, B., Unwin, I. D., and Buss, D. H. (1992a) *First supplement to McCance and Widdowson's The Composition of Foods, 5th edition: Fruit and Nuts*, Royal Society of Chemistry, Cambridge

Holland, B., Welch, A. A., and Buss, D. H. (1992b) *Second supplement to McCance and Widdowson's The Composition of Foods, 5th edition: Vegetable Dishes*, Royal Society of Chemistry, Cambridge

Southgate, D. A. T. (1969) Determination of carbohydrates in food II Unavailable carbohydrates. *J. Sci. Food Agric.* **20**, 331-335

The
Tables

Symbols and abbreviations used in the tables

Symbols

0	None of the nutrient is present
Tr	Trace
N	The nutrient is present in significant quantities but there is no reliable information on the amount
( )	Estimated value

Abbreviations

Trypt	Tryptophan
Satd	Saturated
Monounsatd	Monounsaturated
Polyunsatd	Polyunsaturated

Composition of food per 100g

No. 16-	Food	Description and main data sources	Edible Proportion	Water g	Total Nitrogen g	Protein g	Fat g	Carbohydrate g	Energy value kcal	kJ
1	**Ayr**, raw	Ref 2. Imported frozen from Bangladesh	N	78.1	2.54	15.9	1.3	0	75	318
2	**Bass, Sea**, raw	Analytical and literature sources	N	78.6	3.09	19.3	2.5	0	100	421
3	**Bele**, raw	Literature sources. Imported frozen from Bangladesh	N	79.9	2.45	15.3	0.7	0	67	284
4	**Boal**, raw	Literature sources. Imported frozen from Bangladesh	N	73.0	2.46	15.4	2.7	0	86	362
5	**Bombay duck**	Literature sources. Salted dried fish	0.75	16.7	9.87	61.7	4.0	0	283	1197
6	**Bream, Sea**, raw	Literature sources	N	77.7	2.80	17.5	2.9	0	96	405
7	**Catfish**, raw	Literature sources	N	80.1	2.82	17.6	2.8	0	96	403
8	*steamed*	Middle cuts; flesh only	1.00	73.6	3.43	21.4	3.7	0	119	501
9	*-, weighed with bones*	Calculated from steamed	0.85	62.6	2.92	18.2	3.1	0	101	426
10	**Chinese salted fish**, *steamed*	8 assorted varieties, bones removed	N	47.6	5.42	33.9	2.2	0	155	658
11	**Chital**, raw	Refs 2 and 5. Imported frozen from Bangladesh	N	77.5	2.89	18.1	1.9	0	90	378
12	**Cod**, raw	11 samples from assorted outlets, fillets	0.86[a]	80.8	2.93	18.3	0.7	0	80	337
13	*baked*	Baked in the oven with added butter, fillets; flesh only	1.00	76.6	3.43	21.4	1.2	Tr	96	408
14	*-, weighed with bones and skin*	Calculated from baked	0.85	65.1	2.92	18.2	1.0	Tr	82	347
15	*poached*	Poached in milk, butter and salt added, fillets; flesh only	1.00	77.7	3.35	20.9	1.1	Tr	94	396
16	*-, weighed with bones and skin*	Calculated from poached	0.87	67.6	2.91	18.2	1.0	Tr	81	345

[a] Some fillets contained skin and bones. Values ranged from 0.79 to 1.00

White fish

No. 16-	Food	Starch g	Total sugars g	Dietary fibre Southgate method g	Englyst method g	Fatty acids Satd g	Mono- unsatd g	Poly- unsatd g	Cholest- erol mg
1	**Ayr**, raw	0	0	0	0	N	N	N	N
2	**Bass, Sea**, raw	0	0	0	0	0.4	0.6	0.6	(80)
3	**Bele**, raw	0	0	0	0	N	N	N	N
4	**Boal**, raw	0	0	0	0	N	N	N	N
5	**Bombay duck**	0	0	0	0	N	N	N	N
6	**Bream, Sea**, raw	0	0	0	0	N	N	N	38
7	**Catfish**, raw	0	0	0	0	0.4	0.8	0.8	46
8	*steamed*	0	0	0	0	0.5	1.1	1.1	61
9	*-, weighed with bones*	0	0	0	0	0.4	0.9	0.9	52
10	**Chinese salted fish**, *steamed*	0	0	0	0	N	N	N	N
11	**Chital**, raw	0	0	0	0	N	N	N	N
12	**Cod**, raw	0	0	0	0	0.1	0.1	0.3	46
13	*baked*	0	Tr	0	0	(0.3)	(0.2)	(0.4)	(56)
14	*-, weighed with bones and skin*	0	Tr	0	0	(0.3)	(0.1)	(0.3)	(48)
15	*poached*	0	Tr	0	0	(0.3)	(0.1)	(0.3)	(53)
16	*-, weighed with bones and skin*	0	Tr	0	0	(0.3)	(0.1)	(0.3)	(46)

White fish

Inorganic constituents per 100g food

No. 16-	Food	Na	K	Ca	Mg	P	Fe	Cu	Zn	Cl	Mn	Se	I
						mg						μg	
1	**Ayr, raw**	N	N	380	N	180	0.7	0.06	N	N	N	N	N
2	**Bass, Sea, raw**	(69)	N	130	N	410	2.2	Tr	0.3	N	(0.01)	N	N
3	**Bele, raw**	N	N	370	N	330	1.0	N	N	N	N	N	N
4	**Boal, raw**	N	250	83	N	490	0.8	0.05	0.7	N	0.05	N	N
5	**Bombay duck**	N	N	1390	N	240	19.1	N	N	N	N	N	N
6	**Bream, Sea, raw**	(110)	(270)	40	23	(230)	0.5	0.05	0.3	(120)	0.04	N	N
7	**Catfish, raw**	95	280	20	27	180	0.5	0.05	0.7	110	0.01	N	90
8	steamed	110	320	18	36	240	0.6	0.07	0.9	150	0.01	N	(120)
9	-, weighed with bones	92	270	15	31	200	0.5	0.06	0.8	120	0.01	N	(100)
10	**Chinese salted fish**, steamed	5560	480	69	64	270	1.3	0.24	0.7	8320	0.08	N	N
11	**Chital, raw**	34	120	110	N	210	2.1	0.17	N	N	N	N	N
12	**Cod, raw**	60	340	9	22	180	0.1	0.02	0.4	76	0.01	28	110
13	baked	340	350	11	26	190	0.1	0.02	0.5	520	0.01	34	(130)
14	-, weighed with bones and skin	290	300	N	22	160	0.1	0.02	0.4	440	0.01	29	(110)
15	poached	110	330	11	26	180	0.1	0.02	0.5	150	0.01	33	(120)
16	-, weighed with bones and skin	96	290	1	23	160	0.1	0.02	0.4	130	0.01	29	(100)

No. Food 16-	Retinol µg	Carotene µg	Vitamin D µg	Vitamin E mg	Thiamin mg	Ribo-flavin mg	Niacin mg	Trypt 60 mg	Vitamin B6 mg	Vitamin B12 µg	Folate µg	Panto-thenate mg	Biotin µg	Vitamin C mg
1 Ayr, raw	Tr	Tr	Tr	N	N	N	0.5	3.0	N	N	N	N	N	Tr
2 Bass, Sea, raw	Tr	Tr	Tr	N	N	N	N	3.6	N	(4)	N	N	N	Tr
3 Bele, raw	Tr	Tr	Tr	N	N	N	0.3	2.1	N	2	N	N	N	Tr
4 Boal, raw	N	Tr	N	N	N	N	1.0	2.9	N	5	N	N	N	Tr
5 Bombay duck	Tr	Tr	Tr	N	0.03	0.07	5.7	11.5	0.97	N	N	N	N	Tr
6 Bream, Sea, raw	N	Tr	N	N	0.08	0.10	5.4	3.3	0.46	2	N	0.21	N	Tr
7 Catfish, raw	N	Tr	0.5	2.10	0.19	0.07	2.3	3.3	0.35	2	N	0.57	N	Tr
8 steamed	N	Tr	0.7	2.80	0.23	0.09	2.7	4.0	0.46	3	N	0.60	N	Tr
9 -, weighed with bones	N	Tr	0.6	2.38	0.20	0.08	2.3	3.4	0.39	2	N	0.51	N	Tr
10 Chinese salted fish, steamed	Tr	Tr	Tr	N	Tr	0.11	4.1	6.3	N	5	(8)	N	N	Tr
11 Chital, raw	55	Tr	N	N	0.04	0.07	N	3.4	N	N	N	N	N	Tr
12 Cod, raw	2	Tr	Tr	0.44	0.04	0.05	2.4	3.4	0.18	1	12	0.27	1	Tr
13 baked	2	Tr	Tr	0.59	0.03	0.05	2.3	4.0	0.19	2	12	0.26	1	Tr
14 -, weighed with bones and skin	2	Tr	Tr	0.50	0.03	0.04	1.9	3.4	0.16	1	10	0.22	1	Tr
15 poached	(2)	Tr	Tr	0.61	0.04	0.06	2.8	3.9	0.21	2	14	0.31	1	Tr
16 -, weighed with bones and skin	2	Tr	Tr	0.53	0.03	0.05	2.4	3.4	0.18	1	12	0.27	1	Tr

White fish

Composition of food per 100g

No. 16-	Food	Description and main data sources	Edible Proportion	Water g	Total Nitrogen g	Protein g	Fat g	Carbohydrate g	Energy value kcal	Energy value kJ
17	**Cod**, *steamed*	Middle cuts, flesh only	1.00	79.2	2.98	18.6	0.9	0	83	350
18	-, *weighed with bones and skin*	Calculated from steamed	0.81	64.2	2.41	15.1	0.7	0	67	283
19	frozen, raw	11 samples from assorted supermarkets; steaks	1.00	82.4	2.67	16.7	0.6	0	72	306
20	-, *grilled*	12 samples, grilled with butter and salt added; steaks	1.00	78.0	3.32	20.8	1.3	Tr	95	402
21	in batter, *fried in blended oil*	Samples as fried in retail blend oil, fatty acids calculated	1.00	54.9	2.58	16.1	15.4	11.7	247	1031
22	-, *fried in dripping*	Samples as fried in retail blend oil, fatty acids calculated	1.00	54.9	2.58	16.1	15.4	11.7	247	1031
23	-, *fried in retail blend oil*	24 samples from fish and chip shops, fatty acids calculated	1.00	54.9	2.58	16.1	15.4	11.7	247	1031
24	-, *fried in sunflower oil*	Samples as fried in retail blend oil, fatty acids calculated	1.00	54.9	2.58	16.1	15.4	11.7	247	1031
25	coated in batter, frozen, *baked*	5 samples, different brands; baked for 20 minutes	1.00	57.0	2.05	12.8	11.8	14.3	211	883
26	-, *weighed with bones*	Calculated from baked	0.89	50.7	1.82	11.4	10.5	12.7	188	786
27	coated in crumbs, frozen, *fried in blended oil*	10 samples, 7 brands; shallow fried in blended oil, 5 minutes per side	0.98	55.9	1.98	12.4	14.3	15.2	235	983
28	smoked, raw	Samples from assorted outlets, salted and smoked; flesh only	0.99	78.0	2.93	18.3	0.6	0	79	333
29	-, *poached*	Samples poached in milk, butter added	1.00	73.7	3.46	21.6	1.6	Tr	101	426
30	in parsley sauce, frozen, *boiled*	10 samples, 4 brands; boiled in bag for 20 minutes	1.00	82.1	1.92	12.0	2.8	2.8	84	352

White fish

No. Food 16-	Starch g	Total sugars g	Dietary fibre Southgate method g	Dietary fibre Englyst method g	Fatty acids Satd g	Fatty acids Mono-unsatd g	Fatty acids Poly-unsatd g	Cholesterol mg
17 **Cod**, steamed	0	0	0	0	0.2	0.1	0.4	50
18 -, weighed with bones and skin	0	0	0	0	0.1	0.1	0.3	41
19 frozen, raw	0	0	0	0	0.1	0.1	0.2	39
20 -, grilled	0	Tr	0	0	0.4	0.2	0.3	(49)
21 in batter, fried in blended oil	11.7	Tr	0.5	0.5	1.6	5.5	7.5	N
22 -, fried in dripping	11.7	Tr	0.5	0.5	8.6	5.7	0.4	N
23 -, fried in retail blend oil	11.7	Tr	0.5	0.5	4.1	6.9	3.7	N
24 -, fried in sunflower oil	11.7	Tr	0.5	0.5	1.8	3.1	9.7	N
25 coated in batter, frozen, baked	14.2	0.1	(0.7)	(0.6)	3.6	5.9	1.8	38
26 -, weighed with bones	12.6	0.1	(0.6)	(0.5)	3.2	5.2	1.6	34
27 coated in crumbs, frozen, fried in blended oil	15.0	0.2	(1.1)	(0.4)	(1.5)	(5.2)	(7.0)	N
28 smoked, raw	0	0	0	0	0.1	0.1	0.2	46
29 -, poached	0	Tr	0	0	(0.6)	(0.2)	(0.3)	(53)
30 in parsley sauce, frozen, boiled	2.8	Tr	(0.1)	(0.1)	N	N	N	N

White fish

16-017 to 16-030

Inorganic constituents per 100g food

No. 16-	Food	mg										µg	
		Na	K	Ca	Mg	P	Fe	Cu	Zn	Cl	Mn	Se	I
17	**Cod**, steamed	65	360	10	21	240	0.1	0.02	0.5	120	0.01	30	(110)
18	-, weighed with bones and skin	53	290	8	17	190	0.1	0.02	0.4	97	0.01	24	(92)
19	frozen, raw	71	340	8	22	180	0.1	0.06	0.4	120	0.01	27	(110)
20	-, grilled	91	380	10	26	200	0.1	0.07	0.5	140	0.01	33	(130)
21	in batter, fried in blended oil	160	290	67	25	200	0.5	0.04	0.5	160	0.12	N	N
22	-, fried in dripping	160	290	67	25	200	0.5	0.04	0.5	160	0.12	N	N
23	-, fried in retail blend oil	160	290	67	25	200	0.5	0.04	0.5	160	0.12	N	N
24	-, fried in sunflower oil	160	290	67	25	200	0.5	0.04	0.5	160	0.12	N	N
25	coated in batter, frozen, baked	650	250	42	21	250	0.5	0.08	0.5	800	0.14	15	N
26	-, weighed with bones	570	220	37	19	220	0.4	0.07	0.4	710	0.12	13	N
27	coated in crumbs, frozen, fried in blended oil	480	230	43	19	190	0.4	0.08	0.4	650	0.12	17	N
28	smoked, raw	1170	390	(9)	25	190	(0.1)	(0.02)	0.4	1800	(0.01)	(28)	(110)
29	poached	1200	360	(11)	25	190	(0.1)	(0.02)	0.6	1800	(0.01)	(33)	(120)
30	in parsley sauce, frozen, boiled	260	270	51	19	170	0.1	0.04	0.4	N	0.02	N	N

No. 16-	Food	Retinol µg	Carotene µg	Vitamin D µg	Vitamin E mg	Thiamin mg	Ribo-flavin mg	Niacin mg	Trypt 60 mg	Vitamin B6 mg	Vitamin B12 µg	Folate µg	Panto-thenate mg	Biotin µg	Vitamin C mg
17	Cod, steamed	2	Tr	Tr	0.48	0.04	0.05	2.3	3.5	0.19	2	13	0.23	1	Tr
18	-, weighed with bones and skin	2	Tr	Tr	0.39	0.03	0.04	1.9	2.8	0.15	1	11	0.19	1	Tr
19	frozen, raw	2	Tr	Tr	(0.44)	0.04	0.05	1.6	3.1	(0.18)	1	6	(0.27)	(1)	Tr
20	-, grilled	2	Tr	Tr	(1.00)	0.05	0.06	1.9	3.9	(0.22)	2	10	(0.34)	(1)	Tr
21	in batter, fried in blended oil	N	Tr	Tr	N	0.09	0.07	1.7	3.0	0.13	2	57	0.30	3	Tr
22	-, fried in dripping	N	N	Tr	N	0.09	0.07	1.7	3.0	0.13	2	57	0.30	3	Tr
23	-, fried in retail blend oil	N	Tr	Tr	N	0.09	0.07	1.7	3.0	0.13	2	57	0.30	3	Tr
24	-, fried in sunflower oil	N	Tr	Tr	N	0.09	0.07	1.7	3.0	0.13	2	57	0.30	3	Tr
25	coated in batter, frozen, baked	Tr	N	Tr	N	0.07	0.07	1.1	2.4	0.21	N	8	0.34	2	Tr
26	-, weighed with bones	Tr	N	Tr	N	0.06	0.06	1.0	2.1	0.19	N	7	0.30	2	Tr
27	coated in crumbs, frozen, fried in blended oil	Tr	N	Tr	(3.45)	0.07	0.06	1.2	2.3	0.09	N	6	0.32	3	Tr
28	smoked, raw	(2)	Tr	Tr	(0.44)	0.04	0.05	2.0	3.4	0.17	1	5	(0.27)	(1)	Tr
29	poached	(2)	Tr	Tr	(0.61)	0.05	0.08	2.4	4.0	0.19	2	5	(0.27)	(1)	Tr
30	in parsley sauce, frozen, boiled	Tr	Tr	Tr	N	0.06	0.10	1.1	2.2	0.13	N	17	0.47	2	Tr

Composition of food per 100g

No. 16-	Food	Description and main data sources	Edible Proportion	Water g	Total Nitrogen g	Protein g	Fat g	Carbo- hydrate g	Energy value kcal	kJ
31	Coley, raw	Literature sources and estimation from frozen coley	0.47	80.2	2.93	18.3	1.0	0	82	348
32	steamed	Analytical and calculated values, pieces from tail end; flesh only	1.00	74.8	3.73	23.3	1.3	0	105	444
33	-, weighed with bones and skin	Calculated from steamed	0.85	63.6	3.17	19.8	1.1	0	79	337
34	frozen, raw	10 samples, 5 brands; steaks	1.00	81.7	2.62	16.4	0.9	0	74	312
35	Conger eel, raw	Literature sources	N	73.2	2.96	18.1[a]	4.6	0	114	478
36	grilled	Calculated from raw, steaks; flesh only	1.00	67.8	3.56	21.7[a]	5.5	0	137	574
37	-, weighed with bones and skin	Calculated from grilled	0.52	35.3	1.85	11.3[a]	2.9	0	71	299
38	Dab, raw	Ref 4 and literature sources	N	79.1	2.51	15.7	1.2	0	74	311
39	Dover sole, raw	Ref 4 and literature sources	N	80.0	2.90	18.1	1.8	0	89	374
40	Flounder, raw	Literature sources	N	81.2	2.62	16.4	1.8	0	82	345
41	steamed	Analysis and calculation from raw, whole fish without heads; flesh only	1.00	76.6	3.25	20.3	2.2	0	101	427
42	-, weighed with bones and skin	Calculated from steamed	0.56	42.9	1.82	11.4	1.2	0	57	239
43	Flying fish, raw	Literature sources	0.60	77.5	3.36	21.0	0.3	0	86	366
44	Haddock, raw	12 samples from assorted outlets; fillets	0.83[b]	79.4	3.04	19.0	0.6	0	81	345
45	grilled	Calculated from raw, fillets; flesh only	1.00	73.6	3.89	24.3	0.8	0	104	442
46	-, weighed with bones	Calculated from grilled	0.93	68.4	3.62	22.6	0.7	0	97	411

[a] (Total N - non-protein N) x 6.25

[b] Some fillets contained skin and bones

White fish

No. 16-	Food	Starch g	Total sugars g	Dietary fibre Southgate method g	Englyst method g	Fatty acids Satd g	Mono- unsatd g	Poly- unsatd g	Cholest- erol mg
31	Coley, raw	0	0	0	0	0.1	0.3	0.3	(40)
32	steamed	0	0	0	0	0.2	0.3	0.4	55
33	-, weighed with bones and skin	0	0	0	0	0.1	0.3	0.3	47
34	frozen, raw	0	0	0	0	0.1	0.2	0.3	40
35	Conger eel, raw	0	0	0	0	N	N	N	N
36	grilled	0	0	0	0	N	N	N	N
37	-, weighed with bones and skin	0	0	0	0	N	N	N	N
38	Dab, raw	0	0	0	0	N	N	N	N
39	Dover sole, raw	0	0	0	0	N	N	N	50
40	Flounder, raw	0	0	0	0	(0.4)	(0.2)	(0.7)	48
41	steamed	0	0	0	0	0.4	0.3	0.8	60
42	-, weighed with bones and skin	0	0	0	0	0.2	0.2	0.4	34
43	Flying fish, raw	0	0	0	0	N	N	N	N
44	Haddock, raw	0	0	0	0	0.1	0.1	0.2	36
45	grilled	0	0	0	0	0.2	0.1	0.3	46
46	-, weighed with bones	0	0	0	0	0.1	0.1	0.2	43

Inorganic constituents per 100g food

No. 16-	Food	Na	K	Ca	Mg	P	Fe	Cu	Zn	Cl	Mn	Se	I
						mg						µg	
31	**Coley**, raw	86	360	9	(25)	250	(0.3)	0.05	0.5	(84)	(0.01)	(18)	(36)
32	steamed	97	460	19	31	410	0.6	0.06	0.6	83	(0.01)	(23)	(46)
33	-, weighed with bones and skin	82	390	16	26	350	0.5	0.05	0.5	71	(0.01)	(20)	(39)
34	frozen, raw	170	380	9	25	320	0.3	0.05	0.5	84	0.01	18	36
35	**Conger eel**, raw	50	240	71	20	270	1.3	0.20	0.9	100	0.01	N	N
36	grilled	60	290	85	24	320	1.5	0.24	1.1	120	0.01	N	N
37	-, weighed with bones and skin	31	150	44	12	170	0.8	0.12	0.6	62	0.01	N	N
38	**Dab**, raw	77	350	24	24	N	0.3	0.01	0.5	N	0.04	70	30
39	**Dover sole**, raw	100	310	29	49	200	0.8	0.02	0.4	140	N	23	N
40	**Flounder**, raw	92	300	27	23	190	0.4	0.03	0.5	130	0.04	26	25
41	steamed	120	320	34	25	240	0.5	0.04	0.6	150	0.05	32	31
42	-, weighed with bones and skin	64	180	19	14	130	0.3	0.02	0.3	83	0.03	18	17
43	**Flying fish**, raw	79	460	61	20	230	0.8	Tr	N	90	0.01	N	N
44	**Haddock**, raw	67	360	14	24	200	0.1	0.03	0.4	86	0.01	27	250
45	grilled	86	460	18	31	250	0.1	0.04	0.5	110	0.01	35	320
46	-, weighed with bones	80	430	17	29	230	0.1	0.04	0.5	100	0.01	33	300

No. 16-	Food	Retinol µg	Carotene µg	Vitamin D µg	Vitamin E mg	Thiamin mg	Ribo-flavin mg	Niacin mg	Trypt 60 mg	Vitamin B6 mg	Vitamin B12 µg	Folate µg	Panto-thenate mg	Biotin µg	Vitamin C mg
31	Coley, raw	4	Tr	Tr	0.36	(0.15)	(0.20)	(2.3)	3.4	(0.29)	(3)	N	(0.42)	(3)	Tr
32	steamed	5	Tr	Tr	0.46	(0.19)	(0.27)	(2.9)	4.4	(0.40)	(5)	N	(0.46)	(3)	Tr
33	-, weighed with bones and skin	4	Tr	Tr	0.39	(0.16)	(0.23)	(2.4)	3.7	(0.34)	(4)	N	(0.39)	(3)	Tr
34	frozen, raw	4	Tr	Tr	0.36	0.15	0.20	2.3	3.1	0.29	3	N	0.42	3	Tr
35	Conger eel, raw	N	Tr	N	N	0.06	0.04	4.3	3.4	N	3	N	0.24	N	Tr
36	grilled	N	Tr	N	N	0.06	0.04	4.7	4.1	N	4	N	0.27	N	Tr
37	-, weighed with bones and skin	N	Tr	N	N	0.03	0.02	2.4	2.1	N	2	N	0.14	N	Tr
38	Dab, raw	N	Tr	N	0.40	0.10	0.08	2.3	2.9	0.19	2	5	0.86	1	Tr
39	Dover sole, raw	Tr	Tr	Tr	N	0.06	0.10	3.0	3.4	N	N	N	N	N	Tr
40	Flounder, raw	N	Tr	Tr	(0.36)	0.14	0.21	3.4	3.1	0.25	1	11	N	N	Tr
41	steamed	N	Tr	Tr	(0.45)	0.16	0.26	3.8	3.1	0.31	1	14	N	N	Tr
42	-, weighed with bones and skin	N	Tr	Tr	(0.25)	0.09	0.15	2.1	1.7	0.17	1	8	N	N	Tr
43	Flying fish, raw	Tr	Tr	Tr	N	0.02	0.07	4.5	3.9	N	1	3	N	N	Tr
44	Haddock, raw	Tr	Tr	Tr	0.39	0.04	0.07	4.4	3.6	0.39	1	9	0.26	2	Tr
45	grilled	Tr	Tr	Tr	0.50	0.05	0.08	5.0	3.6	0.45	2	0	0.32	2	Tr
46	-, weighed with bones	Tr	Tr	Tr	0.47	0.05	0.07	4.7	3.3	0.42	2	0	0.30	2	Tr

No. 16-	Food	Description and main data sources	Edible Proportion	Water g	Total Nitrogen g	Protein g	Fat g	Carbo-hydrate g	Energy value kcal	kJ
47	**Haddock,** *poached*	Fish 100g, milk 25g and butter 4g. Calculated from raw, cutlets; flesh only	1.00	75.9	2.82	17.7	4.3	1.1	113	477
48	-, *weighed with bones*	Calculated from poached	1.00	64.5	2.40	15.0	3.7	0.9	96	405
49	*steamed*	12 samples steamed for 20 minutes, fillets; flesh only	1.00	78.3	3.34	20.9	0.6	0	89	378
50	-, *weighed with bones and skin*	Calculated from steamed	0.84[a]	65.8	2.81	17.6	0.5	0	75	317
51	*frozen, raw*	11 samples from assorted supermarkets; steaks	1.00	81.6	2.77	17.3	0.5	0	74	313
52	*in batter, fried in blended oil*	Samples as fried in retail blend oil, fatty acids calculated	1.00	57.6	2.74	17.1	14.0	10.0	232	969
53	-, *fried in dripping*	Samples as fried in retail blend oil, fatty acids calculated	1.00	57.6	2.74	17.1	14.0	10.0	232	969
54	-, *fried in retail blend oil*	20 samples purchased from fish and chip shops, fatty acids calculated	1.00	57.6	2.74	17.1	14.0	10.0	232	969
55	*fried in sunflower oil*	Samples as fried in retail blend oil, fatty acids calculated	1.00	57.6	2.74	17.1	14.0	10.0	232	969
56	*in flour, fried in blended oil*	12 samples, shallow fried in blended oil for 7 minutes per side, fillets; flesh only	1.00	70.6	3.38	21.1	4.1	4.5	138	582
57	-, *weighed with bones and skin*	Calculated from fried in blended oil	0.87	61.4	2.94	18.4	3.6	3.9	120	507
58	*in flour, fried in dripping*	12 samples, shallow fried in dripping for 7 minutes per side, fillets; flesh only	1.00	70.6	3.38	21.1	4.1	4.5	138	582
59	-, *weighed with bones and skin*	Calculated from fried in dripping	0.87	61.4	2.94	18.4	3.6	3.9	120	507

[a] Some fillets contained skin and bones

No. 16-	Food	Starch g	Total sugars g	Dietary fibre Southgate method g	Englyst method g	Fatty acids Satd g	Mono-unsatd g	Poly-unsatd g	Cholesterol mg
47	**Haddock,** *poached*	0	1.1	0	0	2.6	1.1	0.3	43
48	*-, weighed with bones*	0	0.9	0	0	2.2	0.9	0.2	37
49	*steamed*	0	0	0	0	0.1	0.1	0.2	38
50	*-, weighed with bones and skin*	0	0	0	0	0.1	0.1	0.2	32
51	*frozen, raw*	0	0	0	0	0.1	0.1	0.2	32
52	*in batter, fried in blended oil*	10.0	Tr	(0.5)	(0.4)	1.5	5.1	6.9	N
53	*-, fried in dripping*	10.0	Tr	(0.5)	(0.4)	7.8	5.2	0.4	N
54	*-, fried in retail blend oil*	10.0	Tr	(0.5)	(0.4)	3.7	6.3	3.4	N
55	*-, fried in sunflower oil*	10.0	Tr	(0.5)	(0.4)	2.0	3.4	10.7	N
56	*in flour, fried in blended oil*	4.5	Tr	0.2	0.2	0.4	1.5	2.0	N
57	*-, weighed with bones and skin*	3.9	Tr	0.2	0.2	0.4	1.3	1.7	N
58	*in flour, fried in dripping*	4.5	Tr	0.2	0.2	2.3	1.5	0.1	N
59	*-, weighed with bones and skin*	3.9	Tr	0.2	0.2	2.0	1.3	0.1	N

No. 16-	Food	Na	K	Ca	Mg	P	Fe	Cu	Zn	Cl	Mn	Se	I
						mg						µg	
47	**Haddock**, *poached*	99	350	39	24	200	0.1	0.03	0.4	140	0.01	24	230
48	*-, weighed with bones*	84	300	33	20	170	0.1	0.02	0.4	120	0.01	21	(190)
49	*steamed*	73	370	26	24	200	0.1	0.02	0.5	100	0.01	28	(260)
50	*-, weighed with bones and skin*	61	310	22	20	160	0.1	0.02	0.4	84	0.01	24	(220)
51	*frozen, raw*	88	300	10	20	170	0.1	0.02	0.3	110	0.01	24	220
52	*in batter, fried in blended oil*	180	280	180	28	270	0.4	0.04	0.5	170	0.14	21	N
53	*-, fried in dripping*	180	280	180	28	270	0.4	0.04	0.5	170	0.14	21	N
54	*-, fried in retail blend oil*	180	280	180	28	270	0.4	0.04	0.5	170	0.14	21	N
55	*-, fried in sunflower oil*	180	280	180	28	270	0.4	0.04	0.5	170	0.14	21	N
56	*in flour, fried in blended oil*	71	380	49	27	220	0.2	0.01	0.5	96	0.03	28	330
57	*-, weighed with bones and skin*	62	330	43	23	190	0.2	0.01	0.4	84	0.03	24	290
58	*in flour, fried in dripping*	71	380	49	27	220	0.2	0.01	0.5	96	0.03	28	330
59	*-, weighed with bones and skin*	62	330	43	23	190	0.2	0.01	0.4	84	0.03	24	290

No. 16-	Food	Retinol µg	Carotene µg	Vitamin D µg	Vitamin E mg	Thiamin mg	Ribo- flavin mg	Niacin mg	Trypt 60 mg	Vitamin B6 mg	Vitamin B12 µg	Folate µg	Panto- thenate mg	Biotin µg	Vitamin C mg
47	Haddock, poached	43	22	Tr	0.44	0.04	0.10	3.5	3.3	0.36	1	9	0.25	2	Tr
48	-, weighed with bones	36	19	Tr	0.37	0.03	0.09	3.0	2.8	0.31	1	8	0.21	2	Tr
49	steamed	Tr	Tr	Tr	0.41	0.04	0.11	4.1	3.9	0.41	2	9	0.25	1	Tr
50	-, weighed with bones and skin	Tr	Tr	Tr	0.34	0.03	0.09	3.5	3.3	0.34	1	8	0.21	1	Tr
51	frozen, raw	Tr	Tr	Tr	0.35	0.05	0.08	3.1	3.2	(0.39)	(1)	(9)	(0.26)	(2)	Tr
52	in batter, fried in blended oil	N	Tr	Tr	N	0.07	0.13	2.2	3.2	0.23	N	27	0.47	2	Tr
53	-, fried in dripping	N	N	Tr	N	0.07	0.13	2.2	3.2	0.23	N	27	0.47	2	Tr
54	-, fried in retail blend oil	N	Tr	Tr	N	0.07	0.13	2.2	3.2	0.23	N	27	0.47	2	Tr
55	-, fried in sunflower oil	N	Tr	Tr	N	0.07	0.13	2.2	3.2	0.23	N	27	0.47	2	Tr
56	in flour, fried in blended oil	Tr	Tr	Tr	(1.00)	0.04	0.10	4.3	3.9	0.45	2	N	0.24	2	Tr
57	-, weighed with bones and skin	Tr	Tr	Tr	(0.86)	0.03	0.09	3.7	3.4	0.39	2	N	0.21	2	Tr
58	in flour, fried in dripping	N	N	N	(0.40)	0.04	0.10	4.3	3.9	0.45	2	N	0.24	2	Tr
59	-, weighed with bones and skin	N	N	N	(0.35)	0.03	0.09	3.7	3.4	0.39	2	N	0.21	2	Tr

No. 16-	Food	Description and main data sources	Edible Proportion	Water g	Total Nitrogen g	Protein g	Fat g	Carbo-hydrate g	Energy value kcal	Energy value kJ
60	Haddock, in flour, *fried in sunflower oil*	12 samples, shallow fried in sunflower oil for 7 minutes per side, fillets; flesh only	1.00	70.6	3.38	21.1	4.1	4.5	138	582
61	-, *weighed with bones and skin*	Calculated from fried in sunflower oil	0.87	61.4	2.94	18.4	3.6	3.9	120	507
62	coated in crumbs, frozen, raw	10 samples, 7 brands; fillets and steaks	0.81[a]	65.3	2.21	13.8	5.2	10.5	141	595
63	-; *fried in blended oil*	10 samples, 7 brands; shallow fried in blended oil for 10-15 minutes per side	1.00	59.9	2.35	14.7	10.0	12.6	196	822
64	-, *weighed with bones and skin*	Calculated from fried in blended oil	0.80[b]	47.9	1.88	11.8	8.0	10.1	157	657
65	smoked, raw	10 samples from assorted outlets, cutlets	0.84	78.3	3.04	19.0	0.6	0	81	345
66	-, *poached*	Fish 100g, milk 25g and butter 6g. Calculated from raw, cutlets; flesh only	1.00	71.9	2.99	18.7	6.1	1.1	134	562
67	-, *weighed with bones and skin*	Calculated from poached	1.00	64.7	2.69	16.8	5.5	1.0	121	505
68	smoked, *steamed*	Analysis and calculation from raw, cutlets; flesh only	1.00	71.6	3.73	23.3	0.9	0	101	429
69	-, *weighed with bones and skin*	Calculated from steamed	0.65	46.5	2.42	15.1	0.6	0	66	279
70	Hake, raw	Literature sources	N	79.0	2.88	18.0	2.2	0	92	387
71	*grilled*	Calculated from raw, steaks; flesh only	1.00	74.1	3.55	22.2	2.7	0	113	478
72	-, *weighed with bones and skin*	Calculated from grilled	0.65	48.2	2.31	14.4	1.8	0	74	310
73	Halibut, raw	Literature sources	N	76.3	3.44	21.5	1.9	0	103	436
74	*grilled*	Calculated from raw, cutlets and steaks; flesh only	1.00	72.1	4.05	25.3	2.2	0	121	513
75	-, *weighed with bones and skin*	Calculated from grilled	0.78	56.2	3.16	19.7	1.7	0	95	400
76	*poached*	Fish 100g, milk 20g and butter 3g. Calculated from raw, cutlets and steaks; flesh only	1.00	68.1	3.95	24.7	5.7	1.1	154	648

[a] Levels ranged from 0.66 to 1.00

[b] Levels ranged from 0.56 to 1.00

No. 16-	Food	Starch g	Total sugars g	Dietary fibre		Fatty acids			Cholesterol mg
				Southgate method g	Englyst method g	Satd g	Mono-unsatd g	Poly-unsatd g	
60	**Haddock**, in flour, *fried in sunflower oil*	4.5	Tr	0.2	0.2	0.5	0.8	2.6	N
61	-, *weighed with bones and skin*	3.9	Tr	0.2	0.2	0.4	0.7	2.2	N
62	coated in crumbs, frozen, raw	10.5	Tr	(0.6)	(0.6)	N	N	N	N
63	-, *fried in blended oil*	12.6	Tr	(0.6)	(0.6)	N	N	N	N
64	-, *weighed with bones and skin*	10.1	Tr	(0.5)	(0.5)	N	N	N	N
65	smoked, raw	0	0	0	0	(0.1)	(0.1)	(0.2)	(36)
66	-, *poached*	0	1.1	0	0	3.7	1.5	0.4	(50)
67	-, *weighed with bones and skin*	0	1.0	0	0	3.4	1.3	0.3	(45)
68	smoked, *steamed*	0	0	0	0	(0.2)	(0.1)	(0.3)	(47)
69	-, *weighed with bones and skin*	0	0	0	0	(0.1)	(0.1)	(0.2)	(31)
70	**Hake**, raw	0	0	0	0	0.3	0.6	0.5	23
71	grilled	0	0	0	0	0.4	0.7	0.7	28
72	-, *weighed with bones and skin*	0	0	0	0	0.3	0.5	0.4	18
73	**Halibut**, raw	0	0	0	0	0.3	0.6	0.4	35
74	grilled	0	0	0	0	0.4	0.7	0.5	41
75	-, *weighed with bones and skin*	0	0	0	0	0.3	0.5	0.4	32
76	poached	0	1.1	0	0	2.7	1.6	0.6	50

Inorganic constituents per 100g food

No. 16-	Food	Na	K	Ca	Mg	P	Fe	Cu	Zn	Cl	Mn	Se	I
							mg					µg	
60	**Haddock**, in flour, *fried in sunflower oil*	71	380	49	27	220	0.2	0.01	0.5	96	0.03	28	330
61	-, *weighed with bones and skin*	62	330	43	23	190	0.2	0.01	0.4	84	0.03	24	290
62	coated in crumbs, frozen, raw	290	250	120	22	170	0.6	0.05	0.3	400	0.17	22	210
63	-, *fried in blended oil*	290	230	120	21	200	0.8	0.05	0.4	400	0.21	18	250
64	-, *weighed with bones and skin*	230	180	96	17	160	0.6	0.04	0.3	320	0.17	14	200
65	smoked, raw	760	340	22	23	180	0.1	0.03	0.3	1200	0.01	28	(260)
66	-, *poached*	770	350	49	24	190	0.1	0.03	0.4	1220	0.01	27	(250)
67	-, *weighed with bones and skin*	690	310	44	22	170	0.1	0.03	0.3	1100	0.01	24	(230)
68	smoked, steamed	990	440	29	30	240	0.1	0.04	0.4	1570	0.01	36	(340)
69	-, *weighed with bones and skin*	640	280	19	19	150	0.1	0.03	0.3	1020	0.01	23	(220)
70	**Hake**, raw	(100)	(270)	(14)	(23)	(190)	(0.5)	0.03	0.3	(83)	0.02	N	N
71	grilled	(130)	(340)	(17)	(28)	(240)	(0.6)	0.04	0.4	(100)	0.02	N	N
72	-, *weighed with bones and skin*	(84)	(220)	(11)	(18)	(150)	(0.4)	0.03	0.2	(66)	0.01	N	N
73	**Halibut**, raw	60	410	29	25	200	0.5	0.04	0.4	60	0.01	N	40
74	grilled	71	490	34	29	240	0.6	0.05	0.5	71	0.01	N	47
75	-, *weighed with bones and skin*	55	380	27	23	190	0.5	0.04	0.4	55	0.01	N	37
76	poached	100	490	58	30	250	0.6	0.05	0.6	130	0.01	N	47

No. 16-	Food	Retinol µg	Carotene µg	Vitamin D µg	Vitamin E mg	Thiamin mg	Ribo-flavin mg	Niacin mg	Trypt 60 mg	Vitamin B6 mg	Vitamin B12 µg	Folate µg	Panto-thenate mg	Biotin µg	Vitamin C mg
60	**Haddock**, in flour, *fried in sunflower oil*	Tr	Tr	Tr	(2.02)	0.04	0.10	4.3	3.9	0.45	2	N	0.24	2	Tr
61	*-, weighed with bones and skin*	Tr	Tr	Tr	(1.76)	0.03	0.09	3.7	3.4	0.39	2	N	0.21	2	Tr
62	coated in crumbs, frozen, raw	Tr	Tr	Tr	N	0.08	0.06	2.5	2.6	0.31	1	N	0.28	1	Tr
63	*-, fried in blended oil*	Tr	Tr	Tr	N	0.08	0.08	2.8	2.7	0.24	1	N	0.26	2	Tr
64	*-, weighed with bones and skin*	Tr	Tr	Tr	N	0.06	0.06	2.2	2.2	0.19	1	N	0.21	1	Tr
65	smoked, raw	Tr	Tr	Tr	N	0.04	0.12	3.6	3.6	0.35	2	(9)	0.25	1	Tr
66	*-, poached*	61	32	0.1	0.14	0.04	0.15	3.1	3.5	0.34	2	(9)	0.26	1	Tr
67	*-, weighed with bones and skin*	55	28	Tr	0.12	0.04	0.14	2.8	3.2	0.31	1	(8)	0.23	1	Tr
68	smoked, *steamed*	Tr	Tr	Tr	N	0.05	0.16	4.3	4.4	0.46	2	N	0.26	1	Tr
69	*-, weighed with bones and skin*	Tr	Tr	Tr	N	0.03	0.10	2.8	2.8	0.30	1	N	0.17	1	Tr
70	**Hake**, raw	N	Tr	Tr	N	N	N	N	3.4	N	N	N	N	N	Tr
71	*grilled*	N	Tr	Tr	N	N	N	N	4.1	N	N	N	N	N	Tr
72	*-, weighed with bones and skin*	N	Tr	Tr	N	N	N	N	2.7	N	N	N	N	N	Tr
73	**Halibut**, raw	N	Tr	N	0.85	0.07	0.07	5.8	4.0	0.38	1	9	0.31	3	Tr
74	*grilled*	N	Tr	N	1.00	0.07	0.07	6.1	4.7	0.40	1	11	0.35	4	Tr
75	*-, weighed with bones and skin*	N	Tr	N	0.78	0.06	0.06	4.8	3.7	0.31	1	8	0.27	3	Tr
76	*poached*	41	21	N	1.04	0.08	0.12	5.8	4.6	0.44	1	11	0.34	3	Tr

White fish

Composition of food per 100g

No. 16-	Food	Description and main data sources	Edible Proportion	Water g	Total Nitrogen g	Protein g	Fat g	Carbo-hydrate g	Energy value kcal	kJ
77	**Halibut, poached, weighed with bones and skin**	Calculated from poached	1.00	62.6	3.63	22.7	5.3	1.0	142	596
78	**Hoki,** raw	Literature sources. White fleshed fish from New Zealand	N	80.4	2.70	16.9	1.9	0	85	358
79	*grilled*	Calculated from raw, fillets	1.00	72.0	3.86	24.1	2.7	0	121	510
80	**John Dory,** raw	Analytical and literature sources	N	78.1	3.04	19.0	1.4	0	89	375
81	**Kalabasu,** raw	Ref 2. Imported frozen from Bangladesh	N	81.0	2.35	14.7	1.0	0	68	287
82	**Lemon sole,** raw	Literature sources	N	81.2	2.78	17.4	1.5	0	83	351
83	*grilled*	Calculated from raw, whole fish and fillets; flesh only	1.00	78.1	3.23	20.2	1.7	0	97	408
84	*-, weighed with bones and skin*	Calculated from grilled	0.64	50.0	2.07	12.9	1.1	0	62	261
85	*steamed*	Flesh only	1.00	77.2	3.29	20.6	0.9	0	91	384
86	*-, weighed with bones and skin*	Calculated from steamed	0.71	54.8	2.34	14.6	0.6	0	64	272
87	*goujons, baked*	Calculated from manufacturers' proportions	1.00	54.1	2.58	16.0	14.6	14.7	187	775
88	*goujons, fried in blended oil*	Calculated from manufacturers' proportions	1.00	40.8	2.51	15.5	28.7	14.3	374	1553
89	*-, fried in lard*	Calculated from manufacturers' proportions	1.00	41.0	2.51	15.5	28.5	14.3	372	1548
90	*-, fried in sunflower oil*	Calculated from manufacturers' proportions	1.00	40.8	2.51	15.5	28.7	14.3	374	1553
91	**Ling,** raw	Literature sources	N	79.3	3.01	18.8	0.7	0	82	346
92	**Monkfish,** raw	Literature sources	N	83.3	2.51	15.7	0.4	0	66	282
93	*grilled*	Calculated from raw, pieces from tail end; flesh only	1.00	75.8	3.64	22.7	0.6	0	96	407
94	*-, weighed with bones*	Calculated from grilled	0.77	58.4	2.80	17.5	0.4	0	74	314

No. 16-	Food	Starch g	Total sugars g	Dietary fibre Southgate method g	Dietary fibre Englyst method g	Fatty acids Satd g	Fatty acids Mono-unsatd g	Fatty acids Poly-unsatd g	Cholesterol mg
77	**Hallibut**, poached, weighed with bones and skin	0	1.0	0	0	2.5	1.5	0.5	46
78	**Hoki**, raw	0	0	0	0	0.3	0.5	0.4	N
79	grilled	0	0	0	0	0.5	0.7	0.6	N
80	**John Dory**, raw	0	0	0	0	0.3	0.2	0.5	N
81	**Kalabasu**, raw	0	0	0	0	N	N	N	N
82	**Lemon sole**, raw	0	0	0	0	0.2	0.3	0.5	60
83	grilled	0	0	0	0	0.2	0.4	0.6	70
84	-, weighed with bones and skin	0	0	0	0	0.2	0.2	0.4	45
85	steamed	0	0	0	0	0.1	0.2	0.3	73
86	-, weighed with bones and skin	0	0	0	0	0.1	0.1	0.2	52
87	goujons, baked	13.9	0.9	1.0	N	0	0	0	55
88	goujons, fried in blended oil	13.4	0.9	0.9	N	(2.9)	(11.4)	(12.3)	53
89	-, fried in lard	13.4	0.9	0.9	N	(7.4)	(12.6)	(6.7)	67
90	-, fried in sunflower oil	13.4	0.9	0.9	N	(3.2)	(9.1)	(14.5)	53
91	**Ling**, raw	0	0	0	0	N	N	N	N
92	**Monkfish**, raw	0	0	0	0	0.1	0.1	0.1	14
93	grilled	0	0	0	0	0.1	0.1	0.2	20
94	-, weighed with bones	0	0	0	0	0.1	0.1	0.1	16

31

No. 16-	Food	Na	K	Ca	Mg	P	Fe (mg)	Cu	Zn	Cl	Mn	Se (µg)	I
77	**Halibut, poached,** weighed with bones and skin	96	450	54	28	230	0.5	0.04	0.5	120	0.01	N	44
78	**Hoki,** raw	86	380	15	29	190	0.7	0.03	0.4	N	0.02	42	N
79	grilled	120	550	21	41	270	1.0	0.04	0.6	N	0.03	60	N
80	**John Dory,** raw	60	240	40	20	180	N	0.04	0.3	70	N	N	N
81	**Kalabasu,** raw	N	N	320	N	380	0.8	N	N	N	N	60	N
82	**Lemon sole,** raw	(95)	(230)	(17)	(17)	(200)	(0.5)	0.01	0.4	(97)	N	60	N
83	grilled	(110)	(270)	(20)	(20)	(230)	(0.6)	0.01	0.4	(110)	N	70	N
84	-, weighed with bones and skin	(71)	(170)	(13)	(13)	(150)	(0.4)	0.01	0.3	(72)	N	45	N
85	steamed	120	280	21	20	250	0.6	0.01	0.5	120	N	73	N
86	-, weighed with bones and skin	85	200	15	14	180	0.4	0.01	0.3	85	N	52	N
87	goujons, baked	200	230	49	20	200	0.9	0.04	0.5	260	N	N	N
88	goujons, fried in blended oil	190	220	48	19	190	0.9	0.04	0.5	250	N	N	N
89	-, fried in lard	190	220	48	20	190	0.9	0.04	0.5	250	N	N	N
90	-, fried in sunflower oil	190	220	48	19	190	0.9	0.04	0.5	250	N	N	N
91	**Ling,** raw	120	350	25	62	210	0.7	0.02	0.4	N	N	35	N
92	**Monkfish,** raw	18	300	8	21	330	0.3	0.01	0.5	370	0.02	N	N
93	grilled	26	430	12	30	480	0.5	0.01	0.7	540	0.03	N	N
94	-, weighed with bones	20	330	9	23	370	0.4	0.01	0.5	410	0.02	N	N

White fish

Vitamins per 100g food

No. 16-	Food	Retinol µg	Carotene µg	Vitamin D µg	Vitamin E mg	Thiamin mg	Ribo-flavin mg	Niacin mg	Trypt 60 mg	Vitamin B6 mg	Vitamin B12 µg	Folate µg	Panto-thenate mg	Biotin µg	Vitamin C mg
77	**Halibut**, poached, weighed with bones and skin	38	20	N	0.95	0.07	0.11	5.4	4.3	0.40	1	10	0.31	3	Tr
78	**Hoki**, raw	Tr	Tr	Tr	N	0.02	0.05	1.6	3.2	N	N	N	N	N	Tr
79	*grilled*	Tr	Tr	Tr	N	0.03	0.06	2.1	4.5	N	N	N	N	N	Tr
80	**John Dory**, raw	Tr	Tr	Tr	N	N	N	Tr	3.6	0.95	N	N	N	N	Tr
81	**Kalabasu**, raw	Tr	Tr	Tr	N	N	N	0.6	2.7	N	N	N	N	N	Tr
82	**Lemon sole**, raw	Tr	Tr	Tr	N	0.09	0.08	3.5	0.2	N	1	11	0.30	(5)	Tr
83	*grilled*	Tr	Tr	Tr	N	0.10	0.08	3.7	0.3	N	1	13	0.33	(6)	Tr
84	*-, weighed with bones and skin*	Tr	Tr	Tr	N	0.06	0.05	2.3	0.2	N	1	8	0.21	(4)	Tr
85	*steamed*	Tr	Tr	Tr	N	0.10	0.10	3.8	3.8	N	1	13	0.29	5	Tr
86	*-, weighed with bones and skin*	Tr	Tr	Tr	N	0.07	0.07	2.7	2.7	N	1	9	0.21	4	Tr
87	*goujons, baked*	9	3	0	(3.07)	0.08	0.08	2.2	3.0	N	1	11	(0.24)	(4)	Tr
88	*goujons, fried in blended oil*	8	3	Tr	(6.49)	0.09	0.08	2.2	3.0	N	1	14	(0.24)	(3)	Tr
89	*-, fried in lard*	8	3	Tr	(2.99)	0.09	0.08	2.2	3.0	N	1	14	(0.24)	(3)	Tr
90	*-, fried in sunflower oil*	8	3	Tr	(10.14)	0.09	0.08	2.2	3.0	N	1	14	(0.24)	(3)	Tr
91	**Ling**, raw	Tr	Tr	Tr	0.30	0.11	0.08	2.3	3.5	0.30	1	N	0.32	1	Tr
92	**Monkfish**, raw	Tr	Tr	Tr	N	0.03	0.06	N	2.9	N	N	N	N	N	Tr
93	*grilled*	Tr	Tr	Tr	N	0.04	0.08	N	4.3	N	N	N	N	N	Tr
94	*-, weighed with bones*	Tr	Tr	Tr	N	0.03	0.06	N	3.3	N	N	N	N	N	Tr

White fish

Composition of food per 100g

No. 16-	Food	Description and main data sources	Edible Proportion	Water g	Total Nitrogen g	Protein g	Fat g	Carbo-hydrate g	Energy value kcal	Energy value kJ
95	**Mullet, Grey, raw**	Literature sources	0.50	75.2	3.17	19.8	4.0	0	115	485
96	*grilled*	Calculated from raw, whole fish with guts removed; flesh only	1.00	67.8	4.12	25.7	5.2	0	150	629
97	*-, weighed with bones and skin*	Calculated from grilled	0.55	37.3	2.27	14.1	2.9	0	82	346
98	**Mullet, Red, raw**	6 samples, purchased whole	0.23	77.7	2.99	18.7	3.8	0	109	459
99	*grilled*	6 samples, grilled for 7 minutes, whole fish with guts removed; flesh only	1.00	73.9	3.26	20.4	4.4	0	121	510
100	*-, weighed with bones and skin*	Calculated from grilled	0.47	34.7	1.53	9.6	2.1	0	57	240
101	**Parrot fish, raw**	Literature sources	0.65	79.0	3.18	19.9	0.4	0	83	353
102	**Plaice, raw**	8 fish purchased whole, and literature sources	0.42	79.5	2.67	16.7	1.4	0	79	336
103	*grilled*	Calculated from raw, whole fish, heads removed and fillets; flesh only	1.00	75.3	3.22	20.1	1.7	0	96	404
104	*-, weighed with bones and skin*	Calculated from grilled	0.73	55.0	2.35	14.7	1.2	0	70	295
105	*frozen, raw*	12 samples from assorted outlets, fillets	0.77[a]	82.8	2.44	15.2	1.2	0	72	303
106	*-, grilled*	12 samples, grilled 7 minutes per side, fillets; flesh only	1.00	70.9	3.89	25.7	2.0	0	121	511
107	*-, weighed with bones and skin*	Calculated from grilled	0.91[b]	64.5	3.54	23.4	1.8	0	110	465
108	*frozen, steamed*	12 samples, steamed for 15-20 minutes, fillets; flesh only	1.00	78.0	3.14	19.6	1.5[c]	0	92	389
109	*-, weighed with bones and skin*	Calculated from steamed	0.72[d]	56.2	2.26	14.1	1.1	0	66	280

[a] Fillets contained skin and bones
[c] Skin contains 7g fat per 100g

[b] Levels ranged from 0.72 to 1.00
[d] Levels ranged from 0.49 to 0.81

White fish

No. 16-	Food	Starch g	Total sugars g	Dietary fibre Southgate method g	Dietary fibre Englyst method g	Fatty acids Satd g	Fatty acids Mono-unsatd g	Fatty acids Poly-unsatd g	Cholesterol mg
95	**Mullet, Grey,** raw	0	0	0	0	1.1	0.9	0.5	34
96	*grilled*	0	0	0	0	1.4	1.2	0.7	44
97	*-, weighed with bones and skin*	0	0	0	0	0.8	0.6	0.4	24
98	**Mullet, Red,** raw	0	0	0	0	N	N	N	N
99	*grilled*	0	0	0	0	N	N	N	N
100	*-, weighed with bones and skin*	0	0	0	0	N	N	N	N
101	**Parrot fish,** raw	0	0	0	0	(0.1)	Tr	(0.2)	N
102	**Plaice,** raw	0	0	0	0	0.2	0.4	0.3	42
103	*grilled*	0	0	0	0	0.3	0.5	0.4	51
104	*-, weighed with bones and skin*	0	0	0	0	0.2	0.3	0.3	37
105	*frozen, raw*	0	0	0	0	0.2	0.3	0.3	42
106	*-, grilled*	0	0	0	0	0.3	0.6	0.5	71
107	*-, weighed with bones and skin*	0	0	0	0	0.3	0.5	0.5	65
108	*frozen, steamed*	0	0	0	0	0.2	0.4	0.4	54
109	*-, weighed with bones and skin*	0	0	0	0	0.2	0.3	0.3	39

White fish

Inorganic constituents per 100g food

No. 16-	Food	Na	K	Ca	Mg	P	Fe	Cu	Zn	Cl	Mn	Se	I
						mg						µg	
95	**Mullet, Grey**, raw	65	350	27	29	260	1.0	0.07	0.4	130	0.01	(51)	190
96	grilled	84	460	35	38	330	1.3	0.09	0.5	170	0.01	(66)	250
97	-, weighed with bones and skin	46	250	19	21	180	0.7	0.05	0.3	94	0.01	(36)	140
98	**Mullet, Red**, raw	91	340	66	30	220	0.3	0.04	0.3	140	Tr	46	11
99	grilled	110	400	77	35	260	0.4	0.05	0.4	160	Tr	54	13
100	-, weighed with bones and skin	50	190	36	16	120	0.2	0.02	0.2	74	Tr	25	6
101	**Parrot fish**, raw	82	420	45	N	160	1.0	N	N	N	N	N	N
102	**Plaice**, raw	120	280	45	22	180	0.3	0.02	0.5	170	0.01	37	33
103	grilled	140	340	54	27	220	0.4	0.02	0.6	200	0.01	45	40
104	-, weighed with bones and skin	110	250	39	19	160	0.3	0.02	0.4	150	0.01	33	29
105	frozen, raw	84	240	18	18	150	0.2	0.03	0.4	120	0.01	31	28
106	-, grilled	120	320	30	25	220	0.2	0.06	0.8	170	0.03	52	(47)
107	-, weighed with bones and skin	110	290	27	23	200	0.2	0.05	0.7	150	0.03	47	(43)
108	frozen, steamed	110	290	27[a]	22	160	0.2	0.02	0.6	140	0.01	40	(36)
109	-, weighed with bones and skin	76	210	19	16	120	0.1	0.01	0.4	97	0.01	29	(26)

[a] Skin contains 150mg Ca per 100g

No. 16-	Food	Retinol µg	Carotene µg	Vitamin D µg	Vitamin E mg	Thiamin mg	Ribo-flavin mg	Niacin mg	Trypt 60 mg	Vitamin B6 mg	Vitamin B12 µg	Folate µg	Panto-thenate mg	Biotin µg	Vitamin C mg
95	**Mullet, Grey**, raw	47	Tr	N	N	0.06	0.15	3.8	3.7	N	N	N	N	N	Tr
96	grilled	61	Tr	N	N	0.07	0.18	4.4	4.8	N	N	N	N	N	Tr
97	-, weighed with bones and skin	34	Tr	N	N	0.04	0.10	2.4	2.6	N	N	N	N	N	Tr
98	**Mullet, Red**, raw	Tr	Tr	0.8	0.51	0.08	0.08	3.6	3.5	0.42	2	11	0.46	2	Tr
99	grilled	Tr	Tr	0.9	0.59	0.07	0.10	4.2	3.8	0.41	2	10	0.49	3	Tr
100	-, weighed with bones and skin	Tr	Tr	0.4	0.28	0.03	0.05	2.0	1.8	0.19	1	5	0.23	1	Tr
101	**Parrot fish**, raw	N	Tr	N	N	0.06	0.10	1.4	3.7	0.30	3	N	N	N	Tr
102	**Plaice**, raw	Tr	Tr	Tr	N	0.20	0.19	3.2	3.1	0.22	1	11	0.80	(47)	Tr
103	grilled	Tr	Tr	Tr	N	0.22	0.21	3.5	3.7	0.24	2	13	0.92	(57)	Tr
104	-, weighed with bones and skin	Tr	Tr	Tr	N	0.16	0.15	2.5	2.7	0.18	1	10	0.67	(41)	Tr
105	frozen, raw	Tr	Tr	Tr	N	0.20	0.21	2.0	2.9	0.15	1	(11)	0.70	47	Tr
106	-, grilled	Tr	Tr	Tr	N	0.26	0.26	2.8	4.5	0.23	2	(19)	1.13	79	Tr
107	-, weighed with bones and skin	Tr	Tr	Tr	N	0.24	0.24	2.6	4.1	0.21	2	(17)	1.03	72	Tr
108	frozen, steamed	Tr	Tr	Tr	N	0.29	0.17	2.4	3.7	0.24	1	(14)	0.87	48	Tr
109	-, weighed with bones and skin	Tr	Tr	Tr	N	0.21	0.12	1.7	2.6	0.17	1	(10)	0.63	34	Tr

White fish

Composition of food per 100g

No. 16-	Food	Description and main data sources	Edible Proportion	Water g	Total Nitrogen g	Protein g	Fat g	Carbo-hydrate g	Energy value kcal	kJ
110	**Plaice**, in batter, *fried in blended oil*	Samples as fried in retail blend oil, fatty acids calculated	1.00	52.4	2.43	15.2	16.8	12.0	257	1072
111	-, *fried in dripping*	Samples as fried in retail blend oil, fatty acids calculated	1.00	52.4	2.43	15.2	16.8	12.0	257	1072
112	-, *fried in retail blend oil*	20 samples purchased from fish and chip shops, fatty acids calculated	1.00	52.4	2.43	15.2	16.8	12.0	257	1072
113	-, *fried in sunflower oil*	Samples as fried in retail blend oil, fatty acids calculated	1.00	52.4	2.43	15.2	16.8	12.0	257	1072
114	in crumbs, *fried in blended oil*	8 fillets, dipped in egg and breadcrumbs and fried; flesh only, light skin included	1.00	59.9	2.88	18.0	13.7	8.6	228	951
115	-, *fried in lard*	8 fillets, dipped in egg and breadcrumbs and fried; flesh only, light skin included	1.00	59.9	2.88	18.0	13.7	8.6	228	951
116	-, *fried in sunflower oil*	8 fillets, dipped in egg and breadcrumbs and fried; flesh only, light skin included	1.00	59.9	2.88	18.0	13.7	8.6	228	951
117	in crumbs, frozen, *fried in blended oil*	12 samples, 8 brands; shallow fried in blended oil for 5 minutes per side	1.00	64.3	2.27	14.2	9.2	9.2	174	729
118	-, *weighed with bones and skin*	Calculated from fried in blended oil	0.74[a]	47.6	1.68	10.5	6.8	6.8	129	539
119	goujons, baked	Calculated from manufacturers' proportions	1.00	40.8	1.43	8.8	18.3	27.7	304	1270
120	goujons, *fried in blended oil*	Calculated from manufacturers' proportions	1.00	27.9	1.39	8.5	32.3	27.0	426	1771
121	-, *fried in lard*	Calculated from manufacturers' proportions	1.00	28.1	1.39	8.5	32.1	27.0	425	1766
122	-, *fried in sunflower oil*	Calculated from manufacturers' proportions	1.00	27.9	1.39	8.5	32.3	27.0	426	1771

[a] Levels ranged from 0.61 to 0.99

No. 16-	Food	Starch g	Total sugars g	Dietary fibre Southgate method g	Dietary fibre Englyst method g	Fatty acids Satd g	Fatty acids Mono- unsatd g	Fatty acids Poly- unsatd g	Cholest- erol mg
110	**Plaice,** in batter, *fried in*								
	blended oil	12.0	Tr	(0.5)	(0.5)	1.8	6.1	8.2	N
111	-, *fried in dripping*	12.0	Tr	(0.5)	(0.5)	9.4	6.3	0.4	N
112	-, *fried in retail blend oil*	12.0	Tr	(0.5)	(0.5)	4.5	7.6	4.0	N
113	-, *fried in sunflower oil*	12.0	Tr	(0.5)	(0.5)	2.0	3.4	10.7	N
114	in crumbs, *fried in blended oil*	(8.3)	(0.3)	0.4	(0.2)	1.5	4.9	6.7	N
115	-, *fried in lard*	(8.3)	(0.3)	0.4	(0.2)	5.7	6.1	1.3	N
116	-, *fried in sunflower oil*	(8.3)	(0.3)	0.4	(0.2)	1.6	2.8	4.9	N
117	in crumbs, frozen, *fried in*								
	blended oil	9.2	Tr	(0.7)	(0.3)	N	N	N	N
118	-, *weighed with bones and skin*	6.8	Tr	(0.5)	(0.2)	N	N	N	N
119	goujons, *baked*	27.0	0.9	1.5	N	0	0	0	23
120	goujons, *fried in blended oil*	26.1	0.9	1.4	N	(3.4)	(12.9)	(13.3)	22
121	-, *fried in lard*	26.1	0.9	1.4	N	(7.8)	(14.1)	(7.7)	36
122	-, *fried in sunflower oil*	26.1	0.9	1.4	N	(3.6)	(10.6)	(15.5)	22

No. 16-	Food	mg										µg	
		Na	K	Ca	Mg	P	Fe	Cu	Zn	Cl	Mn	Se	I
110	**Plaice**, in batter, fried in blended oil	210	210	73	20	170	0.5	0.06	0.7	240	0.14	6	N
111	-, fried in dripping	210	210	73	20	170	0.5	0.06	0.7	240	0.14	6	N
112	-, fried in retail blend oil	210	210	73	20	170	0.5	0.06	0.7	240	0.14	6	N
113	-, fried in sunflower oil	210	210	73	20	170	0.5	0.06	0.7	240	0.14	6	N
114	in crumbs, fried in blended oil	220	280	67	24	180	0.8	(0.02)	0.7	310	(0.16)	(29)	(31)
115	-, fried in lard	220	280	67	24	180	0.8	(0.02)	0.7	310	(0.16)	(29)	(31)
116	-, fried in sunflower oil	220	280	67	24	180	0.8	(0.02)	0.7	310	(0.16)	(29)	(31)
117	in crumbs, frozen, fried in blended oil	220	210	130	17	140	0.7	0.02	0.3	330	0.16	29	31
118	-, weighed with bones and skin	160	160	93	13	100	0.5	0.01	0.2	240	0.12	21	23
119	goujons, baked	480	220	130	21	120	1.2	0.06	0.5	710	0.28	N	N
120	goujons, fried in blended oil	470	210	120	20	120	1.2	0.06	0.5	690	0.27	N	N
121	-, fried in lard	470	210	120	20	120	1.2	0.06	0.5	690	0.27	N	N
122	-, fried in sunflower oil	470	210	120	20	120	1.2	0.06	0.5	690	0.27	N	N

No. 16-	Food	Retinol µg	Carotene µg	Vitamin D µg	Vitamin E mg	Thiamin mg	Ribo-flavin mg	Niacin mg	Trypt 60 mg	Vitamin B6 mg	Vitamin B12 µg	Folate µg	Panto-thenate mg	Biotin µg	Vitamin C mg
110	**Plaice**, in batter, fried in blended oil	N	Tr	Tr	N	0.19	0.32	1.8	2.8	0.12	N	27	0.82	30	Tr
111	-, fried in dripping	N	N	N	N	0.19	0.32	1.8	2.8	0.12	N	27	0.82	30	Tr
112	-, fried in retail blend oil	N	Tr	Tr	N	0.19	0.32	1.8	2.8	0.12	N	27	0.82	30	Tr
113	-, fried in sunflower oil	N	Tr	Tr	N	0.19	0.32	1.8	2.8	0.12	N	27	0.82	30	Tr
114	in crumbs, fried in blended oil	Tr	Tr	Tr	(3.31)	0.23	0.18	2.9	3.4	(0.15)	1	17	(0.52)	(30)	Tr
115	-, fried in lard	Tr	Tr	Tr	N	0.23	0.18	2.9	3.4	(0.15)	1	17	(0.52)	(30)	Tr
116	-, fried in sunflower oil	Tr	Tr	Tr	(6.75)	0.23	0.18	2.9	3.4	(0.15)	1	17	(0.52)	(30)	Tr
117	in crumbs, frozen, fried in blended oil	Tr	Tr	Tr	N	0.26	0.15	1.7	2.7	0.15	1	(17)	0.52	30	Tr
118	-, weighed with bones and skin	Tr	Tr	Tr	N	0.19	0.11	1.3	2.0	0.11	1	(13)	0.38	22	Tr
119	goujons, baked	9	3	0	(3.42)	0.07	0.11	1.2	1.7	0.20	1	5	(0.29)	N	Tr
120	goujons, fried in blended oil	9	3	Tr	6.83	0.08	0.10	1.2	1.7	0.16	1	6	(0.28)	N	Tr
121	-, fried in lard	9	3	Tr	3.33	0.08	0.10	1.2	1.7	0.16	1	6	(0.28)	N	Tr
122	-, fried in sunflower oil	9	3	Tr	10.48	0.08	0.10	1.2	1.7	0.16	1	6	(0.28)	N	Tr

No. 16-	Food	Description and main data sources	Edible Proportion	Water g	Total Nitrogen g	Protein g	Fat g	Carbohydrate g	Energy value kcal	Energy value kJ
123	Pollack, Alaskan, raw	Literature sources. Used in processed fish foods	0.41	81.4	2.66	16.6	0.6	0	72	304
124	Pomfret, black, raw	Refs 2 and 5	0.70	77.5	3.00	18.8	2.6	0	99	416
125	white, raw	Literature sources	0.68	74.9	2.93	18.3	2.1	0	92	389
126	Red snapper, raw	Literature sources and calculation from fried	0.47	78.3	3.14	19.6	1.3	0	90	381
127	fried in blended oil	10 whole fish, shallow fried for 7 minutes per side; flesh only	1.00	71.4	3.92	24.5	3.1	0	126	531
128	-, weighed with bones and skin	Calculated from fried in blended oil	0.51	36.4	2.00	12.5	1.6	0	64	271
129	fried in sunflower oil	10 whole fish, shallow fried for 7 minutes per side; flesh only	1.00	71.4	3.92	24.5	3.1	0	126	531
130	-, weighed with bones and skin	Calculated from fried in sunflower oil	0.51	36.4	2.00	12.5	1.6	0	64	271
131	Redfish, raw	Literature sources	0.45	78.1	2.94	18.4	2.7	0	98	413
132	Rohu, raw	Ref 2. Imported frozen from Bangladesh	0.78	76.7	2.66	16.6	1.4	0	79	334
133	Rock Salmon/Dogfish, raw	Data from Seafish Industry Authority and literature sources	N	68.3	3.24	16.6[a]	9.7	0	154	641
134	in batter, fried in blended oil	Samples as fried in retail blend oil, fatty acids calculated	1.00	51.3	2.83	14.7[a]	21.9	10.3	295	1225
135	-, weighed with bones	Calculated from fried in blended oil	0.93	47.7	2.63	13.7[a]	20.4	9.6	274	1139
136	in batter, fried in dripping	Samples as fried in retail blend oil, fatty acids calculated	1.00	51.3	2.83	14.7[a]	21.9	10.3	295	1225
137	-, weighed with bones	Calculated from fried in dripping	0.93	47.7	2.63	13.7[a]	20.4	9.6	274	1139

[a] (Total N - non-protein N) x 6.25

White fish

No. 16-	Food	Starch g	Total sugars g	Dietary fibre Southgate method g	Englyst method g	Fatty acids Satd g	Mono- unsatd g	Poly- unsatd g	Cholest- erol mg
123	Pollack, Alaskan, raw	0	0	0	0	(0.1)	(0.1)	(0.2)	71
124	Pomfret, black, raw	0	0	0	0	(0.7)	(0.3)	(0.8)	N
125	white, raw	0	0	0	0	0.6	0.4	0.4	N
126	Red snapper, raw	0	0	0	0	0.3	0.2	0.4	37
127	fried in blended oil	0	0	0	0	0.6	0.8	1.2	46
128	-, weighed with bones and skin	0	0	0	0	0.3	0.4	0.6	23
129	fried in sunflower oil	0	0	0	0	0.7	0.6	1.4	46
130	-, weighed with bones and skin	0	0	0	0	0.3	0.3	0.7	23
131	Redfish, raw	0	0	0	0	0.5	0.8	0.5	42
132	Rohu, raw	0	0	0	0	N	N	N	N
133	Rock Salmon/Dogfish, raw	0	0	0	0	1.4	2.6	2.7	76
134	in batter, fried in blended oil	10.3	Tr	(0.5)	(0.4)	(2.9)	(8.0)	(9.9)	N
135	-, weighed with bones	9.6	Tr	(0.5)	(0.4)	(2.7)	(7.4)	(9.2)	N
136	in batter, fried in dripping	10.3	Tr	(0.5)	(0.4)	(9.6)	(8.1)	(3.1)	N
137	-, weighed with bones	9.6	Tr	(0.5)	(0.4)	(8.9)	(7.6)	(2.9)	N

White fish

Inorganic constituents per 100g food

No. 16-	Food	Na	K	Ca	Mg	P	Fe	Cu	Zn	Cl	Mn	Se	I
						mg						μg	
123	**Pollack**, Alaskan, raw	99	330	9	43	280	0.2	0.04	0.4	N	0.01	35	N
124	**Pomfret**, black, raw	110	430	32	N	250	1.9	N	N	N	N	N	N
125	white, raw	94	320	72	N	300	0.9	0.05	1.3	N	0.18	N	N
126	**Red snapper**, raw	77	370	40	27	210	0.3	0.03	0.3	130	0.01	27	49
127	fried in blended oil	120	460	53	35	270	0.4	0.03	0.4	170	0.02	36	65
128	-, weighed with bones and skin	59	230	27	18	140	0.2	0.02	0.2	87	0.01	18	33
129	fried in sunflower oil	120	460	53	35	270	0.4	0.03	0.4	170	0.02	36	65
130	-, weighed with bones and skin	59	230	27	18	140	0.2	0.02	0.2	87	0.01	18	33
131	**Redfish**, raw	73	300	15	29	180	0.6	Tr	0.4	N	Tr	51	N
132	**Rohu**, raw	100	290	N	13	180	1.0	0.13	N	N	N	N	N
133	**Rock Salmon/Dogfish**, raw	120	290	8	21	230	0.9	0.04	0.4	N	0.02	(55)	N
134	in batter, fried in blended oil	160	230	44	18	190	0.5	0.08	0.4	170	0.14	22	N
135	-, weighed with bones	150	210	41	17	180	0.5	0.07	0.4	160	0.13	20	N
136	in batter, fried in dripping	160	230	44	18	190	0.5	0.08	0.4	170	0.14	22	N
137	-, weighed with bones	150	210	41	17	180	0.5	0.07	0.4	160	0.13	20	N

White fish

No. 16-	Food	Retinol µg	Carotene µg	Vitamin D µg	Vitamin E mg	Thiamin mg	Ribo- flavin mg	Niacin mg	Trypt 60 mg	Vitamin B6 mg	Vitamin B12 µg	Folate µg	Panto- thenate mg	Biotin µg	Vitamin C mg
123	**Pollack**, Alaskan, raw	Tr	Tr	Tr	N	0.07	0.06	1.3	3.1	N	3	3	0.14	N	Tr
124	**Pomfret**, black, raw	Tr	Tr	Tr	N	0.02	0.09	2.1	3.5	N	N	N	N	N	Tr
125	white, raw	Tr	Tr	Tr	N	0.16	0.19	3.7	3.4	0.45	2	N	N	N	Tr
126	**Red snapper**, raw	3	Tr	2.3	N	0.08	0.08	6.3	3.7	0.44	1	N	0.32	1	Tr
127	*fried in blended oil*	4	Tr	3.0	1.35	0.08	0.08	6.6	4.6	0.46	2	N	0.33	1	Tr
128	*-, weighed with bones and skin*	2	Tr	1.5	0.69	0.04	0.04	3.3	2.3	0.23	1	N	0.17	1	Tr
129	*fried in sunflower oil*	4	Tr	3.0	N	0.08	0.08	6.6	4.6	0.46	2	N	0.33	1	Tr
130	*-, weighed with bones and skin*	2	Tr	1.5	N	0.04	0.04	3.3	2.3	0.23	1	N	0.17	1	Tr
131	**Redfish**, raw	8	Tr	N	1.25	0.11	0.09	2.3	3.4	0.18	2	N	0.36	N	Tr
132	**Rohu**, raw	Tr	Tr	Tr	N	0.05	0.07	0.7	3.1	N	5	N	N	N	Tr
133	**Rock Salmon/Dogfish**, raw	190	Tr	(9.1)	N	0.17	0.29	2.9	3.1	0.37	5	3	0.75	N	Tr
134	*in batter, fried in blended oil*	94	Tr	N	N	0.07	0.08	3.2	2.8	0.21	N	4	0.57	12	Tr
135	*-, weighed with bones*	87	Tr	N	N	0.07	0.07	3.0	2.6	0.20	N	4	0.53	11	Tr
136	*in batter, fried in dripping*	N	N	N	N	0.07	0.08	3.2	2.8	0.21	N	4	0.57	12	Tr
137	*-, weighed with bones*	N	N	N	N	0.07	0.07	3.0	2.6	0.20	N	4	0.53	11	Tr

Composition of food per 100g

No. 16-	Food	Description and main data sources	Edible Proportion	Water g	Total Nitrogen g	Protein g	Fat g	Carbohydrate g	Energy value kcal	Energy value kJ
138	**Rock Salmon/Dogfish,** in batter, *fried in retail blend oil*	10 samples purchased from fish and chip shops, fatty acids calculated	1.00	51.3	2.83	14.7[a]	21.9	10.3	295	1225
139	-, *weighed with bones*	Calculated from fried in retail blend oil	0.93	47.7	2.63	13.7[a]	20.4	9.6	274	1139
140	in batter, *fried in sunflower oil*	Samples as fried in retail blend oil, fatty acids calculated	1.00	51.3	2.83	14.7[a]	21.9	10.3	295	1225
141	-, *weighed with bones*	Calculated from fried in sunflower oil	0.93	47.7	2.63	13.7[a]	20.4	9.6	274	1139
142	**Shark,** raw	Data from Seafish Industry Authority and literature sources	N	74.8	3.67	23.0[b]	1.1	0	102	432
143	**Skate,** raw	Data from Seafish Industry Authority and literature sources	N	80.7	3.46	15.1[a]	0.4	0	64	272
144	grilled	Calculated from raw	1.00	75.9	4.32	18.9[a]	0.5	0	79	337
145	-, *weighed with bones*	Calculated from grilled, wings	0.74	56.2	3.20	14.0[a]	0.3	0	59	250
146	in batter, *fried in blended oil*	Samples as fried in retail blend oil, fatty acids calculated	1.00	50.7	3.67	14.7[a]	10.1	4.9	168	702
147	-, *weighed with bones*	Calculated from fried in blended oil	0.85	43.1	3.12	12.5[a]	8.6	4.2	143	597
148	in batter, *fried in dripping*	Samples as fried in retail blend oil, fatty acids calculated	1.00	50.7	3.67	14.7[a]	10.1	4.9	168	702
149	-, *weighed with bones*	Calculated from fried in dripping	0.85	43.1	3.12	12.5[a]	8.6	4.2	143	597

[a] (Total N - non-protein N) x 6.25

[b] Value will include calculated contribution from non-protein nitrogen

No. 16-	Food	Starch g	Total sugars g	Dietary fibre Southgate method g	Englyst method g	Fatty acids Satd g	Mono- unsatd g	Poly- unsatd g	Cholest- erol mg
138	**Rock Salmon/Dogfish,** in batter fried in retail blend oil	10.3	Tr	(0.5)	(0.4)	5.3	9.3	6.2	N
139	-, weighed with bones	9.6	Tr	(0.5)	(0.4)	4.9	8.6	5.8	N
140	in batter, fried in sunflower oil	10.3	Tr	(0.5)	(0.4)	(3.1)	(5.6)	(12.0)	N
141	-, weighed with bones	9.6	Tr	(0.5)	(0.4)	(2.9)	(5.2)	(11.2)	N
142	**Shark,** raw	0	0	0	0	0.2	0.2	0.4	44
143	**Skate,** raw	0	0	0	0	Tr	0.1	0.2	N
144	grilled	0	0	0	0	0.1	0.1	0.2	N
145	-, weighed with bones	0	0	0	0	Tr	0.1	0.1	N
146	in batter, fried in blended oil	(4.8)	(0.1)	0.2	(0.2)	1.0	3.4	4.7	N
147	-, weighed with bones	(4.1)	(0.1)	0.2	(0.2)	0.9	2.9	4.0	N
148	in batter, fried in dripping	(4.8)	(0.1)	0.2	(0.2)	5.3	3.5	0.2	N
149	-, weighed with bones	(4.1)	(0.1)	0.2	(0.2)	4.5	3.0	0.2	N

No. 16-	Food	Na	K	Ca	Mg	P	mg Fe	Cu	Zn	Cl	Mn	µg Se	I
138	**Rock Salmon/Dogfish,** in batter, fried in retail blend oil	160	230	44	18	190	0.5	0.08	0.4	170	0.14	22	N
139	-, weighed with bones	150	210	41	17	180	0.5	0.07	0.4	160	0.13	20	N
140	in batter, fried in sunflower oil	160	230	44	18	190	0.5	0.08	0.4	170	0.14	22	N
141	-, weighed with bones	150	210	41	17	180	0.5	0.07	0.4	160	0.13	20	N
142	**Shark,** raw	140	310	(18)	N	N	0.7	0.05	N	N	0.01	N	N
143	**Skate,** raw	120	260	40	30	180	0.5	0.02	0.5	210	Tr	N	20
144	grilled	150	320	50	37	220	0.6	0.03	0.6	260	Tr	N	25
145	-, weighed with bones	110	240	37	28	160	0.5	0.02	0.4	190	Tr	N	19
146	in batter, fried in blended oil	140	240	50	27	180	1.0	0.09	0.9	170	N	N	N
147	-, weighed with bones	120	200	43	23	150	0.9	0.08	0.8	140	N	N	N
148	in batter, fried in dripping	140	240	50	27	180	1.0	0.09	0.9	170	N	N	N
149	-, weighed with bones	120	200	43	23	150	0.9	0.08	0.8	140	N	N	N

16-138 to 16-149
Vitamins per 100g food

No. 16-	Food	Retinol µg	Carotene µg	Vitamin D µg	Vitamin E mg	Thiamin mg	Ribo- flavin mg	Niacin mg	Trypt 60 mg	Vitamin B6 mg	Vitamin B12 µg	Folate µg	Panto- thenate mg	Biotin µg	Vitamin C mg
138	**Rock Salmon/Dogfish,** in batter, fried in retail blend oil	94	Tr	N	N	0.07	0.08	3.2	2.8	0.21	N	4	0.57	12	Tr
139	-, weighed with bones	87	Tr	N	N	0.07	0.07	3.0	2.6	0.20	N	4	0.53	11	Tr
140	in batter, fried in sunflower oil	94	Tr	N	N	0.07	0.08	3.2	2.8	0.21	N	4	0.57	12	Tr
141	-, weighed with bones	87	Tr	N	N	0.07	0.07	3.0	2.6	0.20	N	4	0.53	11	Tr
142	**Shark,** raw	N	Tr	N	N	0.23	0.30	4.0	4.3	0.42	5	N	N	N	Tr
143	**Skate,** raw	N	Tr	N	N	0.12	0.20	2.0	2.8	0.37	6	N	N	N	Tr
144	grilled	N	Tr	N	N	0.15	0.25	2.5	3.6	0.46	8	N	N	N	Tr
145	-, weighed with bones	N	Tr	N	N	0.11	0.19	1.9	2.6	0.34	6	N	N	N	Tr
146	in batter, fried in blended oil	9	Tr	N	1.20	0.03	0.10	2.4	2.7	N	N	N	N	N	Tr
147	-, weighed with bones	8	Tr	N	1.02	0.03	0.09	2.0	2.3	N	N	N	N	N	Tr
148	in batter, fried in dripping	N	N	N	N	0.03	0.10	2.4	2.7	N	N	N	N	N	Tr
149	-, weighed with bones	N	N	N	N	0.03	0.09	2.0	2.3	N	N	N	N	N	Tr

49

No. 16-	Food	Description and main data sources	Edible Proportion	Water g	Total Nitrogen g	Protein g	Fat g	Carbo-hydrate g	Energy value kcal	Energy value kJ
150	Skate, in batter, fried in retail blend oil	6 samples purchased from fish and chip shops, fatty acids calculated	1.00	50.7	3.67	14.7[a]	10.1	4.9	168	702
151	-, weighed with bones	Calculated from fried in retail blend oil	0.85	43.1	3.12	12.5[a]	8.6	4.2	143	597
152	in batter, fried in sunflower oil	6 samples purchased from fish and chip shops	1.00	50.7	3.67	14.7[a]	10.1	4.9	168	702
153	-, weighed with bones	Calculated from fried in sunflower oil	0.85	43.1	3.12	12.5[a]	8.6	4.2	143	597
154	Tilapia, raw	Literature sources. Type of bream	0.37	79.3	2.85	17.8	1.5	0	85	358
155	Turbot, raw	Literature sources	N	78.7	2.83	17.7	2.7	0	95	401
156	grilled	Calculated from raw, whole fish and fillets; flesh only	1.00	72.7	3.63	22.7	3.5	0	122	514
157	-, weighed with bones and skin	Calculated from grilled	0.62	45.1	2.25	14.1	2.1	0	76	319
158	White fish, dried, salted	Samples purchased whole from West Indian outlets	0.75	44.5	5.53	34.5	1.1	0	148	627
159	Whiting, raw	11 samples from assorted outlets, whole fish and fillets	0.39[b]	80.7	2.99	18.7	0.7	0	81	344
160	steamed	Analysis and calculation from steamed, flesh only	1.00	76.9	3.35	20.9	0.9	0	92	389
161	-, weighed with bones and skin	Calculated from steamed	0.93	71.5	3.12	19.4	0.8	0	85	361
162	in crumbs, fried in blended oil	Fillets coated in crumbs and fried	1.00	63.0	2.90	18.1	10.3	7.0	191	801
163	-, weighed with bones and skin	Calculated from fried in blended oil	0.90	56.7	2.61	16.3	9.3	6.3	172	721
164	in crumbs, fried in dripping	Fillets coated in crumbs and fried	1.00	63.0	2.90	18.1	10.3	7.0	191	801
165	-, weighed with bones and skin	Calculated from fried in dripping	0.90	56.7	2.61	16.3	9.3	6.3	172	721
166	in crumbs, fried in sunflower oil	Fillets coated in crumbs and fried	1.00	63.0	2.90	18.1	10.3	7.0	191	801
167	-, weighed with bones and skin	Calculated from fried in sunflower oil	0.90	56.7	2.61	16.3	9.3	6.3	172	721

[a] (Total N - non-protein N) x 6.25

[b] This value is for whole whiting. The edible proportion for whiting fillets is 0.73

No. Food 16-	Starch g	Total sugars g	Dietary fibre Southgate method g	Dietary fibre Englyst method g	Fatty acids Satd g	Fatty acids Mono- unsatd g	Fatty acids Poly- unsatd g	Cholest- erol mg
150 **Skate**, in batter, *fried in retail blend oil*	(4.8)	(0.1)	0.2	(0.2)	2.5	4.3	2.3	N
151 -, *weighed with bones*	(4.1)	(0.1)	0.2	(0.2)	2.1	3.6	1.9	N
152 *in batter, fried in sunflower oil*	(4.8)	(0.1)	0.2	(0.2)	1.1	1.9	6.0	N
153 -, *weighed with bones*	(4.1)	(0.1)	0.2	(0.2)	1.0	1.6	5.1	N
154 **Tilapia**, raw	0	0	0	0	0.3	0.4	0.3	N
155 **Turbot**, raw	0	0	0	0	0.7	0.6	0.6	N
156 *grilled*	0	0	0	0	0.9	0.8	0.7	N
157 -, *weighed with bones and skin*	0	0	0	0	0.5	0.5	0.5	N
158 **White fish**, dried, salted	0	0	0	0	N	N	N	N
159 **Whiting**, raw	0	0	0	0	0.1	0.2	0.2	46
160 *steamed*	0	0	0	0	0.1	0.3	0.2	55
161 -, *weighed with bones and skin*	0	0	0	0	0.1	0.2	0.2	51
162 *in crumbs, fried in blended oil*	6.8	0.2	0.3	0.2	1.1	3.7	5.0	N
163 -, *weighed with bones and skin*	6.1	0.2	0.3	0.2	1.0	3.3	4.5	N
164 *in crumbs, fried in dripping*	6.8	0.2	0.3	0.2	5.7	3.8	0.3	N
165 -, *weighed with bones and skin*	6.1	0.2	0.3	0.2	5.2	3.5	0.2	N
166 *in crumbs, fried in sunflower oil*	6.8	0.2	0.3	0.2	1.2	2.1	6.5	N
167 -, *weighed with bones and skin*	6.1	0.2	0.3	0.2	1.1	1.9	5.9	N

White fish

Inorganic constituents per 100g food

No. 16-	Food	Na	K	Ca	Mg	P	Fe	Cu	Zn	Cl	Mn	Se	I
							mg					µg	
150	**Skate**, in batter, fried in retail blend oil	140	240	50	27	180	1.0	0.09	0.9	170	N	N	N
151	-, weighed with bones	120	200	43	23	150	0.9	0.08	0.8	140	N	N	N
152	in batter, fried in sunflower oil	140	240	50	27	180	1.0	0.09	0.9	170	N	N	N
153	-, weighed with bones	120	200	43	23	150	0.9	0.08	0.8	140	N	N	N
154	**Tilapia**, raw	(52)	(450)	120	N	350	(1.9)	N	N	N	N	N	N
155	**Turbot**, raw	68	260	49	48	200	0.5	0.04	0.2	140	N	N	N
156	grilled	9	340	62	62	260	0.6	0.05	0.3	180	N	N	N
157	-, weighed with bones and skin	6	210	38	38	160	0.4	0.03	0.2	110	N	N	N
158	**White fish**, dried, salted	7530	300	140	26	210	0.8	0.12	1.1	11000	0.02	21	67
159	**Whiting**, raw	90	330	18	22	170	0.1	0.01	0.4	120	0.01	25	80
160	steamed	110	400	21	28	190	0.1	0.01	0.4	140	0.01	23	74
161	-, weighed with bones and skin	100	370	20	26	180	0.1	0.01	0.4	130	0.01	N	N
162	in crumbs, fried in blended oil	200	320	48	33	260	0.7	N	N	190	N	N	N
163	-, weighed with bones and skin	180	290	43	30	230	0.6	N	N	170	N	N	N
164	in crumbs, fried in dripping	200	320	48	33	260	0.7	N	N	190	N	N	N
165	-, weighed with bones and skin	180	290	43	30	230	0.6	N	N	170	N	N	N
166	in crumbs, fried in sunflower oil	200	320	48	33	260	0.7	N	N	190	N	N	N
167	-, weighed with bones and skin	180	290	43	30	230	0.6	N	N	170	N	N	N

White fish

No. 16-	Food	Retinol μg	Carotene μg	Vitamin D μg	Vitamin E mg	Thiamin mg	Ribo-flavin mg	Niacin mg	Trypt 60 mg	Vitamin B6 mg	Vitamin B12 μg	Folate μg	Panto-thenate mg	Biotin μg	Vitamin C mg
150	**Skate**, in batter, fried in retail blend oil	9	Tr	N	1.20	0.03	0.10	2.4	2.7	N	N	N	N	N	Tr
151	-, weighed with bones	8	Tr	N	1.02	0.03	0.09	2.0	2.3	N	N	N	N	N	Tr
152	in batter, fried in sunflower oil	9	Tr	N	N	0.03	0.10	2.4	2.7	N	N	N	N	N	Tr
153	-, weighed with bones	8	Tr	N	N	0.03	0.09	2.0	2.3	N	N	N	N	N	Tr
154	**Tilapia**, raw	N	Tr	Tr	N	(0.03)	(0.09)	(3.1)	3.3	(0.32)	(2)	N	N	N	Tr
155	**Turbot**, raw	Tr	Tr	Tr	N	0.05	0.11	2.1	3.3	N	2	N	0.57	N	Tr
156	grilled	Tr	Tr	Tr	N	0.06	0.14	2.4	4.2	2.52	0	N	0.66	N	Tr
157	-, weighed with bones and skin	Tr	Tr	Tr	N	0.04	0.09	1.5	2.6	1.56	0	N	0.41	N	Tr
158	**White fish**, dried, salted	N	Tr	N	N	0.07	0.07	2.7	6.5	N	3	3	N	N	Tr
159	**Whiting**, raw	Tr	Tr	Tr	N	0.05	0.26	1.7	3.5	0.17	N	N	0.25	1	Tr
160	steamed	Tr	Tr	Tr	N	0.05	0.31	1.8	3.9	0.20	N	N	0.24	1	Tr
161	-, weighed with bones and skin	Tr	Tr	Tr	N	0.05	0.29	1.7	3.6	0.19	N	N	0.22	1	Tr
162	in crumbs, fried in blended oil	Tr	Tr	Tr	(2.48)	N	N	N	3.4	N	N	N	N	N	Tr
163	-, weighed with bones and skin	Tr	Tr	Tr	(2.24)	N	N	N	3.0	N	N	N	N	N	Tr
164	in crumbs, fried in dripping	N	N	Tr	N	N	N	N	3.4	N	N	N	N	N	Tr
165	-, weighed with bones and skin	N	N	Tr	N	N	N	N	3.0	N	N	N	N	N	Tr
166	in crumbs, fried in sunflower oil	Tr	Tr	Tr	(5.07)	N	N	N	3.4	N	N	N	N	N	Tr
167	-, weighed with bones and skin	Tr	Tr	Tr	(4.57)	N	N	N	3.0	N	N	N	N	N	Tr

Fatty fish

Composition of food per 100g

No. 16-	Food	Description and main data sources	Edible Proportion	Water g	Total Nitrogen g	Protein g	Fat g	Carbo-hydrate g	Energy value kcal	kJ
168	Anchovies, canned in oil, drained	10 samples, 4 brands	0.74	41.6	4.03	25.2	19.9	0	280	1165
169	Bacha, raw	Ref 2. Imported frozen from Bangladesh	N	68.8	2.90	18.1	5.6	0	123	515
170	Bloater, grilled	Salted and smoked herring; flesh only	1.00	55.6	3.76	23.5	17.4	0	251	1043
171	-, weighed with bones and skin	Calculated from grilled	0.65	36.1	2.44	15.3	11.3	0	163	678
172	Carp, raw	Literature sources	0.54	77.3	2.80	17.5	4.7	0	112	471
173	Eel, raw	Analytical and literature sources. Yellow eels	0.67	71.3[a]	2.66	16.6	11.3[b]	0	168	700
174	jellied	10 samples from assorted outlets. Jelly included in analysis	0.95	82.7	1.35	8.4	7.1	Tr	98	406
175	Herring, raw	35 fish, purchased whole over the year	0.50	68.0[c]	2.85	17.8	13.2[d]	0	190	791
176	grilled	Samples gutted, grilled for 7 minutes per side; flesh only	1.00	63.9	3.22	20.1	11.2	0	181	756
177	-, weighed with bones and skin	Calculated from grilled	0.68	43.5	2.19	13.7	7.6	0	123	514
178	in oatmeal, fried in vegetable oil	Samples shallow fried; flesh, skin and roes	1.00	58.7	3.69	23.1	15.1	1.5	234	975
179	-, weighed with bones	Calculated from fried in vegetable oil	0.77	45.2	2.84	17.8	11.6	1.2	180	751
180	dried, salted	Samples purchased whole from West Indian outlets, flesh only	1.00	49.7	4.05	25.3	7.4	0	168	704
181	-, weighed with bones and skin	Calculated from dried	0.50	24.9	2.03	12.7	3.7	0	84	352
182	canned in tomato sauce	10 samples, 4 brands; whole contents	1.00	67.4	2.05	12.8	14.4	3.2	193	802
183	pickled	6 samples, loose rollmops and in jars	0.57[e]	57.9	2.67	16.7	11.1	(10.0)	209[f]	877[f]
184	Hilsa, raw	Ref 2. Imported frozen from Bangladesh	N	53.7	3.49	21.8	19.4	0	262	1088

[a] The water content varies according to stage of maturity: elvers contain 81.8g and silver eels 57.1g water per 100g
[b] The fat content varies according to stage of maturity: elvers contain 2.2g and silver eels 27.8g fat per 100g
[c] Levels range from 57g to 79g per 100g being highest in spring and lowest in autumn/winter
[d] Levels range from 5g per 100g in spring to 20g per 100g in winter
[e] Value for drained jars
[f] Includes contribution from acetic acid

54

Fatty fish

No. 16-	Food	Starch g	Total sugars g	Dietary fibre Southgate method g	Dietary fibre Englyst method g	Fatty acids Satd g	Fatty acids Mono-unsatd g	Fatty acids Poly-unsatd g	Cholesterol mg
168	**Anchovies,** canned in oil, drained	0	0	0	0	N	N	N	N
169	**Bacha,** raw	0	0	0	0	N	N	N	N
170	**Bloater,** grilled	0	0	0	0	(4.4)	(7.3)	(3.6)	66
171	-, weighed with bones and skin	0	0	0	0	(2.9)	(4.8)	(2.3)	43
172	**Carp,** raw	0	0	0	0	0.9	1.6	0.8	67
173	**Eel,** raw	0	0	0	0	2.9	5.4	1.6	150
174	jellied	Tr	0	0	0	1.9	3.5	1.0	79
175	**Herring,** raw	0	0	0	0	3.3	5.5	2.7	50
176	grilled	0	0	0	0	2.8	4.7	2.3	43
177	-, weighed with bones and skin	0	0	0	0	1.9	3.2	1.6	29
178	in oatmeal, fried in vegetable oil	1.5	Tr	0.1	0.1	N	N	N	(57)
179	-, weighed with bones	1.2	Tr	0.1	0.1	N	N	N	(44)
180	dried, salted	0	0	0	0	N	N	N	N
181	-, weighed with bones and skin	0	0	0	0	N	N	N	N
182	canned in tomato sauce	0.3	2.9	Tr	Tr	N	N	N	46
183	pickled	0	(10.0)	0	0	N	N	N	(42)
184	**Hilsa,** raw	0	0	0	0	N	N	N	N

Inorganic constituents per 100g food

No. 16-	Food	Na	K	Ca	Mg	P	Fe	Cu	Zn	Cl	Mn	Se (µg)	I (µg)
						mg						µg	
168	**Anchovies**, canned in oil, *drained*	3930	230	300	56	300	4.1	0.17	3.0	6090	0.18	N	N
169	**Bacha**, raw	N	N	520	N	180	0.7	0.11	N	N	N	N	N
170	**Bloater**, grilled	700	450	86	45	360	1.7	0.21	1.3	1130	0.05	50	N
171	-, *weighed with bones and skin*	450	290	56	29	230	1.1	0.14	0.8	730	0.03	32	N
172	**Carp**, raw	43	360	47	36	240	0.9	0.06	1.0	37	0.04	(28)	N
173	**Eel**, raw	89	270	19	19	310	1.2	0.07	2.5	57	0.17	31	(80)
174	jellied	660	55	62	6	73	0.1	0.04	0.9	980	0.04	22	14
175	**Herring**, raw	120	320	60	32	230	1.2	0.14	0.9	170	0.04	35	29
176	grilled	160	430	79	42	310	1.6	0.19	1.2	220	0.05	46	38
177	-, *weighed with bones and skin*	110	300	54	29	210	1.1	0.13	0.8	150	0.03	31	26
178	in oatmeal, *fried in vegetable oil*	160	420	79	35	340	1.6	0.18	1.2	220	0.05	46	38
179	-, *weighed with bones*	120	320	61	27	260	1.2	0.14	0.9	170	0.04	35	29
180	dried, salted	5210	150	200	73	250	1.9	4.10	1.9	7620	0.08	N	N
181	-, *weighed with bones and skin*	2610	75	100	37	130	1.0	2.05	1.0	3810	0.04	N	N
182	canned in tomato sauce	380	320	45	31	170	0.8	0.12	1.5	420	0.06	24	18
183	pickled	830	62	13	5	77	0.7	0.12	0.3	1660	0.02	N	N
184	**Hilsa**, raw	52	180	180	N	280	2.1	0.14	N	N	N	N	N

No. 16-	Food	Retinol μg	Carotene μg	Vitamin D μg	Vitamin E mg	Thiamin mg	Ribo-flavin mg	Niacin mg	Trypt 60 mg	Vitamin B6 mg	Vitamin B12 μg	Folate μg	Panto-thenate mg	Biotin μg	Vitamin C mg
168	Anchovies, canned in oil, drained	57	Tr	N	N	Tr	0.10	3.8	4.7	N	11	18	N	N	Tr
169	Bacha, raw	N	Tr	N	N	N	N	0.6	3.4	N	N	N	N	N	Tr
170	Bloater, grilled	(37)	Tr	(25.0)	(1.00)	Tr	(0.27)	(4.0)	4.4	(0.35)	(15)	(10)	(0.78)	(7)	Tr
171	-, weighed with bones and skin	(24)	Tr	(16.2)	(0.65)	Tr	(0.18)	(2.6)	2.9	(0.23)	(1)	(6)	(0.51)	(4)	Tr
172	Carp, raw	N	Tr	N	0.63	0.09	0.07	2.5	3.3	0.17	2	N	0.15	N	Tr
173	Eel, raw	1200[a]	Tr	4.9[b]	(4.14)	0.20	0.35	2.3	3.1	0.25	1	12	0.14	2	Tr
174	jellied	110	Tr	3.0	2.60	0.07	0.16	0.8	1.6	0.03	2	N	0.35	7	Tr
175	Herring, raw	44	Tr	19.0[c]	0.76	0.01	0.26	4.1	3.3	0.44	13	9	0.81	7	Tr
176	grilled	34	Tr	16.1	0.64	Tr	0.27	4.0	3.8	0.35	15	10	0.78	7	Tr
177	-, weighed with bones and skin	23	Tr	10.9	0.44	Tr	0.18	2.7	2.6	0.24	10	7	0.53	5	Tr
178	in oatmeal, fried in vegetable oil	(50)	Tr	(21.7)	N	Tr	(0.27)	(4.0)	4.3	(0.35)	(15)	(10)	(0.78)	(7)	Tr
179	-, weighed with bones	(38)	Tr	(16.7)	N	Tr	(0.21)	(3.1)	3.3	(0.27)	(11)	(8)	(0.60)	(5)	Tr
180	Herring, dried, salted	N	Tr	N	N	Tr	0.11	3.0	4.7	N	7	15	N	N	Tr
181	-, weighed with bones and skin	N	Tr	N	N	Tr	0.06	1.5	2.4	N	3	8	N	N	Tr
182	canned in tomato sauce	N	285	N	3.57	N	N	N	2.4	N	N	N	N	N	Tr
183	pickled	(37)	Tr	(16.0)	(0.64)	Tr	0.13	0.8	3.1	0.11	N	1	N	N	Tr
184	Hilsa, raw	N	Tr	N	N	N	N	2.8	4.1	N	4	N	N	N	N

[a] Levels range from 260mg to 2500μg retinol per 100g. The content increases with maturity
[b] Whole body oil is a rich source of vitamin D containing about 120μg per 100g oil
[c] Levels range from 7μg to 31μg vitamin D per 100g

Composition of food per 100g

No. 16-	Food	Description and main data sources	Edible Proportion	Water g	Total Nitrogen g	Protein g	Fat g	Carbo-hydrate g	Energy value kcal	kJ
185	**Jackfish**, raw	Data from Seafish Industry Authority and literature sources	0.56	71.9	3.47	21.7	2.4	0	108	458
186	**Katla**, raw	Ref 2 and literature sources. Imported frozen from Bangladesh	N	73.7	3.12	19.5	2.4	0	100	420
187	**Kipper**, raw	10 samples from assorted outlets. Smoked cured herrings	0.55	61.2	2.80	17.5	17.7[a]	0	229	952
188	grilled	10 samples, grilled 4-5 minutes; flesh only	1.00	55.9	3.22	20.1	19.4	0	255	1060
189	-, weighed with bones	Calculated from grilled	0.63	35.2	2.03	12.7	12.2	0	161	667
190	boil in bag, boiled	11 frozen samples with added butter, boiled 12-20 minutes	0.87	60.7	3.20	20.0	17.4	Tr	237	984
191	**Mackerel**, raw	10 samples from assorted outlets, purchased whole; flesh and skin	0.71	64.0[b]	2.99	18.7	16.1[c]	0	220	914
192	fried in blended oil	Analysis and calculation from raw; fried in blended oil; flesh only	1.00	(55.0)	3.83	24.0	19.5	0	272	1130
193	-, weighed with bones and skin	Calculated fried in blended oil	0.73	(40.2)	2.80	17.5	14.2	0	198	825
194	grilled	10 samples, grilled for 5 minutes per side; flesh and skin	1.00	58.6	3.33	20.8	17.3	0	239	994
195	-, weighed with bones and skin	Calculated from grilled	0.92	53.9	3.06	19.1	15.9	0	220	914
196	smoked	10 samples, flesh and skin	0.99	47.1	3.02	18.9	30.9	0	354	1465
197	canned in brine, drained	13 samples, 4 brands; steaks and fillets	0.72	60.5	3.04	19.0	17.9	0	237	985
198	canned in tomato sauce	10 samples, 4 brands; whole contents, fillets	1.00	67.9	2.62	16.4	15.0	1.4	206	856

[a] Levels range from 13.3g to 22.2g fat per 100g
[b] Levels range from 56 to 74g water per 100g
[c] Levels range from 6g to 23g fat per 100g

Fatty fish

No. 16-	Food	Starch g	Total sugars g	Dietary fibre Southgate method g	Englyst method g	Fatty acids Satd g	Mono-unsatd g	Poly-unsatd g	Cholesterol mg
185	**Jackfish**, raw	0	0	0	0	N	N	N	54
186	**Katla**, raw	0	0	0	0	N	N	N	N
187	**Kipper**, raw	0	0	0	0	2.8	9.3	3.9	64
188	*grilled*	0	0	0	0	3.1	10.2	4.2	70
189	*-, weighed with bones*	0	0	0	0	2.0	6.4	2.7	44
190	*boil in bag, boiled*	Tr	Tr	0	0	N	N	N	N
191	**Mackerel**, raw	0	0	0	0	3.3	7.9	3.3	54
192	*fried in blended oil*	0	0	0	0	(4.0)	(9.5)	(4.0)	(69)
193	*-, weighed with bones and skin*	0	0	0	0	(2.9)	(7.0)	(2.9)	(50)
194	*grilled*	0	0	0	0	3.5	8.5	3.5	58
195	*-, weighed with bones and skin*	0	0	0	0	3.3	7.8	3.2	53
196	*smoked*	0	0	0	0	(6.3)	(15.1)	(6.3)	105
197	canned in brine, *drained*	0	0	0	0	(4.0)	(6.8)	(4.6)	60
198	canned in tomato sauce	Tr	1.4	Tr	Tr	(3.3)	(5.7)	(3.8)	56

Fatty fish

Inorganic constituents per 100g food

No. 16-	Food	Na	K	Ca	Mg	P	Fe (mg)	Cu	Zn	Cl	Mn	Se (µg)	I
185	**Jackfish**, raw	53	360	34	31	220	1.6	0.07	0.6	N	0.01	N	30
186	**Katla**, raw	50	150	530	N	240	0.9	0.12	N	N	N	N	N
187	**Kipper**, raw	830	340	53	27	230	1.6	0.12	1.0	1190	0.04	32	55
188	grilled	940	390	60	31	260	1.8	0.14	1.1	1350	0.05	36	63
189	-, weighed with bones	590	240	38	20	160	1.1	0.09	0.7	850	0.03	23	40
190	boil in bag, boiled	700	360	87	39	270	1.2	0.16	1.5	1020	0.06	(33)	(57)
191	**Mackerel**, raw	63	290	11	24	200	0.8	0.08	0.6	82	0.02	30	140
192	fried in blended oil	81	370	14	31	250	1.0	0.10	0.8	110	0.03	38	(180)
193	-, weighed with bones and skin	59	270	10	23	180	0.8	0.07	0.6	77	0.02	28	(130)
194	grilled	63	360	12	28	230	0.8	0.09	0.7	97	0.02	36	(170)
195	-, weighed with bones and skin	58	330	11	26	210	0.7	0.08	0.6	89	0.02	33	(150)
196	smoked	750	310	20	28	210	1.2	0.09	1.1	1130	0.02	33	(150)
197	canned in brine, drained	270	110	29	13	90	1.1	0.28	2.7	610	0.09	42	49
198	canned in tomato sauce	250	310	82	24	190	1.0	0.12	1.1	200	0.06	29	47

Fatty fish

No. 16-	Food	Retinol µg	Carotene µg	Vitamin D µg	Vitamin E mg	Thiamin mg	Ribo-flavin mg	Niacin mg	Trypt 60 mg	Vitamin B6 mg	Vitamin B12 µg	Folate µg	Panto-thenate mg	Biotin µg	Vitamin C mg
185	**Jackfish**, raw	65	Tr	13.0	N	0.17	0.19	6.3	4.1	0.43	5	N	N	N	Tr
186	**Katla**, raw	Tr	Tr	Tr	N	N	N	0.8	3.6	N	1	N	N	N	Tr
187	**Kipper**, raw	32	Tr	8.0	0.32	Tr	0.28	4.1	3.3	0.27	10	6	0.53	5	Tr
188	*grilled*	38	Tr	9.4	0.37	Tr	0.27	4.5	3.8	0.25	12	5	0.51	5	Tr
189	*-, weighed with bones*	24	Tr	5.9	0.23	Tr	0.17	2.8	2.4	0.16	7	3	0.32	3	Tr
190	*boil in bag, boiled*	N	N	(7.9)	(0.31)	Tr	0.41	5.9	3.7	0.26	17	N	1.03	8	Tr
191	**Mackerel**, raw	45	Tr	5.0	0.43	0.14	0.29	8.6	3.5	0.41	8	N	0.81	5	Tr
192	*fried in blended oil*	43	Tr	6.4	N	0.14	0.30	8.8	4.5	0.42	11	N	0.83	5	Tr
193	*-, weighed with bones and skin*	31	Tr	4.7	N	0.10	0.22	6.4	3.3	0.31	8	N	0.61	4	Tr
194	*grilled*	48	Tr	5.4	0.46	0.15	0.32	9.4	3.9	0.45	1	N	0.93	6	Tr
195	*-, weighed with bones and skin*	44	Tr	5.0	0.42	0.14	0.29	8.6	3.6	0.41	9	N	0.86	5	Tr
196	*smoked*	31	Tr	8.0	0.25	0.26	0.52	9.5	3.5	0.50	6	N	1.03	3	Tr
197	*canned in brine, drained*	37	Tr	(5.6)	N	0.03	0.27	5.3	3.6	0.14	8	N	0.59	5	Tr
198	*canned in tomato sauce*	31	165	(4.7)	1.94	0.07	0.20	5.0	3.1	0.20	(8)	6	0.75	6	Tr

Fatty fish

Composition of food per 100g

No. 16-	Food	Description and main data sources	Edible Proportion	Water g	Total Nitrogen g	Protein g	Fat g	Carbohydrate g	Energy value kcal	kJ
199	**Orange roughy**, raw	Literature sources	N	75.9	2.35	14.7	7.0	0	126	527
200	**Pangas**, raw	Ref 2. Imported frozen from Bangladesh	N	72.3	2.27	14.2	10.8	0	154	641
201	**Pilchards**, canned in tomato sauce	10 samples, 6 brands; whole contents	1.00	70.4	2.67	16.7	8.1	1.1	144	601
202	**Salmon**, raw	11 farmed and wild samples, whole fish and steaks	0.79[a]	67.2	3.23	20.2	11.0[b]	0	180	750
203	*grilled*	Calculated from raw, steaks; flesh only	1.00	60.7	3.87	24.2	13.1	0	215	896
204	*-, weighed with bones and skin*	Calculated from grilled	0.82	49.8	3.17	19.8	10.7	0	176	735
205	*steamed*	Calculated from raw, steaks; flesh only	1.00	64.5	3.49	21.8	11.9	0	194	812
206	*-, weighed with bones and skin*	Calculated from steamed	0.77	49.7	2.69	16.8	9.2	0	150	625
207	*smoked*	4 samples	1.00	64.9	4.06	25.4	4.5	0	142	598
208	**Salmon, pink**, canned in brine, flesh only, *drained*	6 samples, 3 brands	0.79	71.3	3.79	23.5	6.6	0	153	644
209	*-, flesh and bones, drained*	6 samples, 3 brands	0.81	71.3	3.79	23.5	6.6	0	153	644
210	**Salmon, red**, canned in brine, flesh only, *drained*	6 samples, 3 brands	0.81	66.2	3.46	21.6	9.0	0	167	700
211	*-, flesh and bones, drained*	6 samples, 3 brands	0.83	71.3	3.79	23.5	6.6	0	153	644
212	**Sardines**, raw	10 samples from assorted outlets, purchased whole	0.47	67.7	3.30	20.6	9.2[c]	0	165	691

[a] This is an average value. The value for whole salmon is 0.62 and salmon steaks 0.81

[b] Wild salmon entering the river contain approximately 14.5g fat per 100g.

[c] Levels range from 5.1g to 13.9g fat per 100g

No. 16-	Food	Starch g	Total sugars g	Dietary fibre Southgate method g	Englyst method g	Fatty acids Satd g	Mono- unsatd g	Poly- unsatd g	Cholest- erol mg
199	Orange roughy, raw	0	0	0	0	0.3	5.9	0.1	20
200	Pangas, raw	0	0	0	0	N	N	N	N
201	Pilchards, canned in tomato sauce	0.2	0.9	Tr	Tr	1.7	2.2	3.4	56
202	Salmon, raw	0	0	0	0	1.9	4.4	3.1	50
203	grilled	0	0	0	0	2.5	5.8	4.1	60
204	-, weighed with bones and skin	0	0	0	0	2.1	4.8	3.4	49
205	steamed	0	0	0	0	2.0	4.7	3.3	54
206	-, weighed with bones and skin	0	0	0	0	1.6	3.6	2.6	42
207	smoked	0	0	0	0	(0.8)	(1.8)	(1.3)	35
208	Salmon, pink, canned in brine, flesh only, drained	0	0	0	0	1.3	2.4	1.9	28
209	-, flesh and bones, drained	0	0	0	0	1.3	2.4	1.9	28
210	Salmon, red, canned in brine, flesh only, drained	0	0	0	0	1.7	3.7	2.4	38
211	-, flesh and bones, drained	0	0	0	0	1.3	2.7	1.7	28
212	Sardines, raw	0	0	0	0	2.7	2.5	2.7	N

Inorganic constituents per 100g food

No. 16-	Food	Na	K	Ca	Mg	P	Fe	Cu	Zn	Cl	Mn	Se	I
						mg						µg	
199	**Orange roughy**, raw	77	330	12	21	150	0.3	0.05	0.3	N	0.01	65	N
200	**Pangas**, raw	N	N	180	N	130	0.5	0.05	N	N	N	N	N
201	**Pilchards**, canned in tomato sauce	290	310	250	29	280	2.5	0.16	1.3	520	0.11	30	64
202	**Salmon**, raw	45	360	21	27	250	0.4	0.03	0.6	58	0.02	(26)	37
203	grilled	54	430	25	32	300	0.5	0.04	0.7	69	0.02	(31)	44
204	-, weighed with bones and skin	44	350	21	27	250	0.4	0.03	0.6	57	0.02	(26)	36
205	steamed	49	390	23	29	270	0.4	0.03	0.7	63	0.02	(28)	4
206	-, weighed with bones and skin	37	300	17	22	210	0.3	0.02	0.5	48	0.02	(22)	3
207	smoked	1880	420	19	32	250	0.6	0.09	0.4	2850	0.02	(24)	N
208	**Salmon, pink**, canned in brine, flesh only, *drained*	430	260	91	25	170	0.6	0.05	0.8	730	Tr	25	59
209	-, flesh and bones, *drained*	440	260	300	33	290	0.8	0.05	1.0	730	0.02	24	76
210	**Salmon, red**, canned in brine, flesh only, *drained*	430	260	91	25	170	0.6	0.05	0.8	730	Tr	25	59
211	-, flesh and bones, *drained*	440	260	300	33	290	0.8	0.05	1.0	730	0.02	24	76
212	**Sardines**, raw	120	360	84	31	270	1.4	0.12	1.0	130	0.06	34	29

No. 16-	Food	Retinol µg	Carotene µg	Vitamin D µg	Vitamin E mg	Thiamin mg	Ribo-flavin mg	Niacin mg	Trypt 60 mg	Vitamin B6 mg	Vitamin B12 µg	Folate µg	Panto-thenate mg	Biotin µg	Vitamin C mg
199	**Orange roughy**, raw	N	Tr	N	N	N	N	N	2.7	N	N	N	N	N	Tr
200	**Pangas**, raw	N	Tr	N	N	N	N	0.6	2.6	N	N	N	N	N	Tr
201	**Pilchards**, canned in tomato sauce	7	(140)	14.0	2.56	0.01	0.33	5.9	3.1	0.27	13	N	0.85	11	Tr
202	**Salmon**, raw	13[a]	Tr	8.0[a]	1.91	0.23	0.13	7.2	3.8	0.75	4	16	1.02	7	Tr
203	grilled	16	Tr	9.6	2.29	0.25	0.14	7.7	4.5	0.81	5	19	1.16	9	Tr
204	-, weighed with bones and skin	13	Tr	7.9	1.88	0.20	0.11	6.3	3.7	0.66	4	16	0.95	7	Tr
205	steamed	14	Tr	8.7	2.07	0.22	0.14	7.0	4.1	0.81	4	17	0.88	7	Tr
206	-, weighed with bones and skin	11	Tr	6.7	1.59	0.17	0.11	5.4	3.1	0.62	3	13	0.68	6	Tr
207	smoked	N	Tr	N	N	0.16	0.17	8.8	4.7	(0.28)	(3)	(2)	(0.87)	N	Tr
208	**Salmon, pink**, canned in brine, flesh only, drained	31	Tr	17.0	1.52	0.02	0.22	5.9	4.4	0.21	4	14	0.74	9	Tr
209	-, flesh and bones, drained	31	Tr	17.0	1.52	0.02	0.22	5.9	4.4	0.21	4	14	0.74	9	Tr
210	**Salmon, red**, canned in brine, flesh only, drained	52	Tr	23.1	2.08	0.02	0.22	5.9	4.0	0.24	4	12	0.74	9	Tr
211	-, flesh and bones, drained	31	Tr	17.0	1.52	0.02	0.22	5.9	4.4	0.21	4	14	0.74	9	Tr
212	**Sardines**, raw	N	Tr	11.0	0.29	Tr	0.22	6.7	3.9	0.39	11	4	0.81	6	Tr

[a] These are values for Atlantic salmon. Pacific salmon may contain 90 (20-150)µg retinol and 12.5 (5-20)µg vitamin D per 100g

No. 16-	Food	Description and main data sources	Edible Proportion	Water g	Total Nitrogen g	Protein g	Fat g	Carbohydrate g	Energy value kcal	kJ
213	**Sardines**, *grilled*	10 samples, grilled whole 5-7 minutes per side; flesh and skin	1.00	64.0	4.05	25.3	10.4	0	195	815
214	-, *weighed with bones*	Calculated from grilled	0.61	39.0	2.47	15.4	6.3	0	119	497
215	canned in brine, *drained*	10 samples, 4 brands	0.79	66.2	3.44	21.5	9.6	0	172	721
216	canned in oil, *drained*	13 samples, 10 brands; canned in vegetable and olive oil	0.82	58.6	3.73	23.3	14.1[a]	0	220	918
217	canned in tomato sauce	10 samples, 8 brands; whole contents	1.00	69.3	2.72	17.0	9.9	1.4	162	678
218	**Sprats**, raw	10 samples from assorted outlets, purchased whole	0.68	66.3	2.93	18.3	11.0	0	172	718
219	*fried*	Analysis and calculation from raw; fish without heads, deep fried	1.00	36.6	3.98	24.9	35.0	0	415	1718
220	-, *weighed with bones*	Calculated from fried	0.60	22.0	2.39	14.9	21.0	0	249	1031
221	**Swordfish**, raw	Data from the Seafish Industry Authority and literature sources	N	80.8	2.88	18.0	4.1	0	109	458
222	*grilled*	Calculated from raw, steaks; flesh only	1.00	75.5	3.67	22.9	5.2	0	139	583
223	-, *weighed with bones and skin*	Calculated from grilled	0.89	67.2	3.27	20.4	4.6	0	123	519
224	**Trout, brown**, raw	Analytical and literature sources	N	74.1	3.10	19.4	3.8	0	112	470
225	**Trout, rainbow**, raw	11 samples from assorted outlets, purchased whole	0.52	76.7	3.14	19.6	5.2[b]	0	125	526
226	*grilled*	11 samples, grilled 7 minutes per side; flesh only	1.00	73.3	3.44	21.5	5.4[b]	0	135	565
227	-, *weighed with bones and skin*	Calculated from grilled	0.73	53.5	2.51	15.7	3.9	0	98	413

[a] If not drained the fat content is approximately 24.4g per 100g

[b] Skin contains 17.3g fat per 100g

Fatty fish

Composition of food per 100g

No. 16-	Food	Starch g	Total sugars g	Dietary fibre Southgate method g	Dietary fibre Englyst method g	Fatty acids Satd g	Fatty acids Mono-unsatd g	Fatty acids Poly-unsatd g	Cholesterol mg
213	**Sardines,** *grilled*	0	0	0	0	3.0	2.9	3.1	N
214	*-, weighed with bones*	0	0	0	0	1.8	1.8	1.9	N
215	canned in brine, *drained*	0	0	0	0	N	N	N	60
216	canned in oil, *drained*	0	0	0	0	2.9	4.8	5.0	65
217	canned in tomato sauce	Tr	1.4	Tr	Tr	2.8	2.9	3.2	76
218	**Sprats,** raw	0	0	0	0	2.2	4.7	3.0	93
219	*fried*	0	0	0	0	(5.3)	(14.0)	(14.2)	125
220	*-, weighed with bones*	0	0	0	0	(3.2)	(8.4)	(8.5)	74
221	**Swordfish,** raw	0	0	0	0	0.9	1.6	1.1	41
222	*grilled*	0	0	0	0	1.2	2.1	1.4	52
223	*-, weighed with bones and skin*	0	0	0	0	1.1	1.9	1.2	46
224	**Trout, brown,** raw	0	0	0	0	N	N	N	N
225	**Trout, rainbow,** raw	0	0	0	0	1.1	1.8	1.7	67
226	*grilled*	0	0	0	0	1.1	2.0	1.7	70[a]
227	*-, weighed with bones and skin*	0	0	0	0	0.8	1.5	1.3	51

[a] Skin contains 230mg cholesterol per 100g

Fatty fish

Inorganic constituents per 100g food

No. 16-	Food	mg										µg	
		Na	K	Ca	Mg	P	Fe	Cu	Zn	Cl	Mn	Se	I
213	**Sardines,** *grilled*	140	400	130	34	320	1.7	0.14	1.4	170	0.08	38	32
214	-, *weighed with bones*	85	240	76	21	200	1.0	0.09	0.9	100	0.05	23	20
215	*canned in brine, drained*	530	320	540	45	510	2.7	0.16	2.3	810	0.20	41	23
216	*canned in oil, drained*	450	410	500	46	520	3.1	0.11	2.2	620	0.19	49	23
217	*canned in tomato sauce*	350	410	430	39	420	2.9	0.16	2.4	590	0.24	37	N
218	**Sprats,** *raw*	200	320	97	37	240	1.1	0.09	1.7	270	0.04	(10)	64
219	*fried*	240	410	120	46	300	1.4	0.11	2.1	340	0.05	(13)	80
220	-, *weighed with bones*	140	250	72	28	180	0.8	0.07	1.3	200	0.03	(8)	48
221	**Swordfish,** *raw*	130	350	4	27	260	0.5	N	N	130	0.02	(45)	N
222	*grilled*	170	450	5	34	340	0.6	N	N	170	0.03	(57)	N
223	-, *weighed with bones and skin*	150	400	5	31	300	0.6	N	N	150	0.02	(51)	N
224	**Trout, brown,** *raw*	56	480	9	29	260	1.5	0.04	1.2	N	0.02	(25)	N
225	**Trout, rainbow,** *raw*	45	420	18	27	240	0.3	0.04	0.5	57	0.01	18	13
226	*grilled*	55	410	21[a]	26	250[a]	0.4	0.05	0.5[a]	65	0.01	21	(15)
227	-, *weighed with bones and skin*	40	300	15	19	180	0.3	0.04	0.4	47	0.01	15	(11)

[a] Skin contains 890mg Ca, 750mg P and 4.1mg Zn per 100g

No. 16-	Food	Retinol µg	Carotene µg	Vitamin D µg	Vitamin E mg	Thiamin mg	Ribo-flavin mg	Niacin mg	Trypt 60 mg	Vitamin B6 mg	Vitamin B12 µg	Folate µg	Panto-thenate mg	Biotin µg	Vitamin C mg
213	**Sardines**, *grilled*	N	Tr	12.3	0.32	Tr	0.25	6.9	4.7	0.41	12	4	0.88	6	Tr
214	-, *weighed with bones*	N	Tr	7.5	0.20	Tr	0.15	4.2	2.9	0.25	7	2	0.54	4	Tr
215	canned in brine, *drained*	6	Tr	4.6	N	0.01	0.26	6.1	4.0	0.16	13	8	0.76	10	Tr
216	canned in oil, *drained*	7	Tr	5.0	0.31	0.01	0.29	6.9	4.4	0.18	15	8	0.86	5	Tr
217	canned in tomato sauce	9	140	8.0	3.08	0.02	0.28	5.5	3.2	0.35	14	13	0.50	5	Tr
218	**Sprats**, *raw*	60	Tr	13.0	0.51	Tr	0.20	3.0	3.4	0.27	7	N	0.71	4	Tr
219	*fried*	74	Tr	N	(5.58)	Tr	0.20	3.0	4.6	0.27	9	N	0.71	5	Tr
220	-, *weighed with bones*	44	Tr	N	(3.35)	Tr	0.12	1.8	2.8	0.16	5	N	0.43	3	Tr
221	**Swordfish**, *raw*	N	Tr	N	N	0.16	0.17	8.3	3.4	0.51	4	N	0.41	N	Tr
222	*grilled*	N	Tr	N	N	0.19	0.20	9.5	4.3	0.59	5	N	0.50	N	Tr
223	-, *weighed with bones and skin*	N	Tr	N	N	0.17	0.17	8.5	3.8	0.52	5	N	0.44	N	Tr
224	**Trout, brown**, *raw*	N	Tr	N	N	N	N	N	3.6	N	N	N	N	N	Tr
225	**Trout, rainbow**, *raw*	49	Tr	10.6	0.71	0.20	0.11	4.5	3.7	0.34	5	9	1.54	3	Tr
226	*grilled*	29	Tr	11.0[a]	1.01[a]	0.20	0.12	4.2	4.0	0.35	5	10	1.58	3	Tr
227	-, *weighed with bones and skin*	21	Tr	8.0	0.74	0.15	0.09	3.1	2.9	0.26	4	7	1.15	2	Tr

[a] Skin contains 2.9mg vitamin E and 24µg vitamin D per 100g

16-228 to 16-231

Composition of food per 100g

No. Food 16-	Description and main data sources	Edible Proportion	Total Water g	Nitrogen g	Protein g	Fat g	Carbo- hydrate g	Energy value kcal	Energy value kJ
228 **Tuna,** raw	Data from Seafish Industry Authority and literature sources	0.58	70.4	3.79	23.7	4.6[a]	0	136	573
229 canned in brine, *drained*	10 samples, 9 brands; skipjack tuna	0.81	74.6	3.76	23.5	0.6	0	99	422
230 canned in oil, *drained*	10 samples, 6 brands; skipjack tuna	0.79	63.3	4.34	27.1	9.0	0	189	794
231 **Whitebait,** in flour, *fried*	Whole fish; rolled in flour and fried	1.00	23.5	3.12	19.5	47.5	5.3	525	2174

[a] Levels can range from 1.0g to 8.8g fat per 100g

Fatty fish

Composition of food per 100g

No. Food 16-	Starch g	Total sugars g	Dietary fibre Southgate method g	Dietary fibre Englyst method g	Fatty acids Satd g	Fatty acids Mono- unsatd g	Fatty acids Poly- unsatd g	Cholest- erol mg
228 **Tuna**, raw	0	0	0	0	1.2	1.2	1.6	28
229 canned in brine, *drained*	0	0	0	0	0.2	0.1	0.2	51
230 canned in oil, *drained*	0	0	0	0	1.5	2.3	4.8	50
231 **Whitebait**, in flour, *fried*	(5.2)	(0.1)	0.2	0.2	N	N	N	N

Fatty fish

Inorganic constituents per 100g food

No. Food	Na	K	Ca	Mg	P	Fe	Cu	Zn	Cl	Mn	Se	I
16-					mg						μg	
228 **Tuna**, raw	47	400	16	33	230	1.3[a]	0.15	0.7	N	0.02	(57)	30
229 canned in brine, *drained*	320	230	8	27	170	1.0	0.05	0.7	550	Tr	78	13
230 canned in oil, *drained*	290	260	12	33	200	1.6	0.20	1.1	530	0.05	90	14
231 **Whitebait**, in flour, *fried*	230	110	860	50	860	5.1	N	N	330	N	N	N

[a]Value is for light tissue, dark tissue contains 9mg Fe per 100g

Fatty fish

No. Food 16-	Retinol µg	Carotene µg	Vitamin D µg	Vitamin E mg	Thiamin mg	Ribo- flavin mg	Niacin mg	Trypt 60 mg	Vitamin B6 mg	Vitamin B12 µg	Folate µg	Panto- thenate mg	Biotin µg	Vitamin C mg
228 **Tuna**, raw	26	Tr	7.2	N	0.10	0.13	12.8	4.4	0.38	4	15	0.66	N	Tr
229 canned in brine, *drained*	N	Tr	4.0	0.55	0.02	0.11	14.4	4.4	0.47	4	4	0.29	2	Tr
230 canned in oil, *drained*	N	Tr	3.0	1.94	0.02	0.12	16.1	5.1	0.51	5	5	0.32	3	Tr
231 **Whitebait**, in flour, *fried*	N	Tr	N	N	N	N	N	3.6	N	N	N	N	N	Tr

73

Nutrient variability

It is important to appreciate that samples of the same or similar fish, as with other foods, will vary somewhat in composition. Some of the differences that occur within a species can be at least as great as differences between species. The amounts of fat and individual fatty acids in fatty fish are particularly variable, although the values in this supplement reflect those that are found in the seasons in which the fish are normally caught. When the fat rises, the water content tends to fall, such that the two together normally provide about 80 per cent of the flesh.

The amount of protein varies much less. There will also be some differences in composition with the length and conditions of storage and with cooking conditions. It is not practical to give separate nutrient values for all these factors, so the tables have been designed to show the most typical values for each product as it is available and eaten in this country.

A more comprehensive description of the factors to be taken into account in the proper use of food composition tables is given in the introduction to the fifth edition of *The Composition of Foods*. Users of the present supplement are advised to read them and take them to heart.

Crustacea

No. 16-	Food	Description and main data sources	Edible Proportion	Water g	Total Nitrogen g	Protein g	Fat g	Carbohydrate g	Energy value kcal	Energy value kJ
232	**Crab**, *boiled*	12 samples; purchased boiled. Light and dark meat	1.00	71.0	3.12	19.5	5.5	Tr	128	535
233	–, *weighed with shell*	Calculated from boiled	0.35	24.8	1.09	6.8	1.9	Tr	45	187
234	canned in brine, *drained*	6 cans, 2 brands. White meat only	0.44	79.2	2.90	18.1	0.5	Tr	77	326
235	**Crayfish**, raw	Literature sources	N	83.8	2.38	14.9	0.8	0	67	283
236	**Lobster**, *boiled*	Boiled in fresh water	1.00	74.3	3.54	22.1	1.6	Tr	103	435
237	–, *weighed with shell*	Calculated from boiled	0.36	26.7	1.27	8.0	0.6	Tr	37	157
238	**Prawns**, raw	Literature sources	0.46	79.2	2.82	17.6	0.6	0	76	321
239	*boiled*	Samples cooked in sea or salt water	1.00	70.0	3.62	22.6	0.9	0	99	418
240	–, *weighed with shells*	Calculated from boiled	0.38	26.6	1.38	8.6	0.3	0	37	159
241	frozen, raw	13 samples, 10 brands	0.54	79.0	2.93	18.3	0.6	0	79	333
242	dried	Ref 5	1.00	13.7	9.98	62.4	3.5	0	281	1190
243	**Scampi**, in breadcrumbs, frozen, *fried in blended oil*	10 samples, 8 brands. Deep fried for 4 minutes	1.00	49.8	1.50	9.4	13.6	20.5	237	991
244	in breadcrumbs, frozen, *fried in sunflower oil*	10 samples, 8 brands. Deep fried for 4 minutes	1.00	49.8	1.50	9.4	13.6	20.5	237	991
245	**Shrimps**, *boiled*	Purchased cooked, probably in sea or salt water	1.00	62.5	3.80	23.8	2.4	Tr	117	493
246	–, *weighed with shells*	Calculated from boiled	0.33	20.6	1.25	7.9	0.8	Tr	39	163
247	canned in brine, *drained*	10 cans, 3 brands	0.65	74.9	3.33	20.8	1.2	Tr	94	398
248	frozen	10 packets from Chinese supermarkets. Shrimps and prawns	1.00	81.2	2.64	16.5	0.8	Tr	73	310
249	dried	5 packets. Shrimps and prawns	1.00	25.5	8.93	55.8	2.4	Tr	245	1037

No. 16-	Food	Starch g	Total sugars g	Dietary fibre		Fatty acids			Cholest- erol mg
				Southgate method g	Englyst method g	Satd g	Mono- unsatd g	Poly- unsatd g	
232	**Crab,** *boiled*	Tr	Tr	0	0	0.7	1.5	1.6	72
233	-, *weighed with shell*	Tr	Tr	0	0	0.2	0.5	0.6	25
234	canned in brine, *drained*	Tr	Tr	0	0	(0.1)	(0.1)	(0.1)	(72)
235	**Crayfish,** raw	0	0	0	0	0.1	0.2	0.3	105
236	**Lobster,** *boiled*	Tr	Tr	0	0	0.2	0.3	0.6	110
237	-, *weighed with shell*	Tr	Tr	0	0	0.1	0.1	0.2	40
238	**Prawns,** raw	0	0	0	0	0.1	0.2	0.1	(195)
239	*boiled*	0	0	0	0	0.2	0.2	0.2	(280)
240	-, *weighed with shells*	0	0	0	0	0.1	0.1	0.1	(105)
241	**Prawns,** frozen, raw	0	0	0	0	0.1	0.2	0.1	195
242	dried	0	0	0	0	(0.8)	(0.9)	(0.8)	N
243	**Scampi,** in breadcrumbs, frozen, *fried in blended oil*	20.5	Tr	1.1	N	1.4	5.1	6.4	110
244	in breadcrumbs, frozen, *fried in sunflower oil*	20.5	Tr	1.1	N	1.6	3.1	8.2	110
245	**Shrimps,** *boiled*	Tr	Tr	0	0	0.4	0.5	0.8	(130)
246	-, *weighed with shells*	Tr	Tr	0	0	0.1	0.2	0.3	(42)
247	canned in brine, *drained*	Tr	Tr	0	0	0.2	0.3	0.4	(130)
248	frozen	Tr	Tr	0	0	(0.1)	(0.2)	(0.3)	(130)
249	dried	Tr	Tr	0	0	(0.4)	(0.5)	(0.8)	(505)

Crustacea

Inorganic constituents per 100g food

No. 16-	Food	Na	K	Ca	Mg	P	Fe (mg)	Cu	Zn	Cl	Mn	Se (µg)	I
232	Crab, *boiled*	420	250	N	58	340	1.6	1.77	5.5	640	0.17	(17)	N
233	-, *weighed with shell*	150	87	N	20	120	0.6	0.62	1.9	220	0.06	(6)	N
234	canned in brine, *drained*	550	100	120	32	140	2.8	0.42	5.7	830	N	N	N
235	Crayfish, raw	150	260	33	25	240	2.2	0.44	1.3	280	(0.15)	(70)	(100)
236	Lobster, *boiled*	330	260	62	34	260	0.8	1.35	2.5	530	(0.03)	(130)	(100)
237	-, *weighed with shell*	120	94	22	12	93	0.3	0.49	0.9	190	(0.01)	(47)	(36)
238	Prawns, raw	190	330	79	(34)	180	1.6	(0.14)	1.5	N	(0.01)	(16)	(21)
239	*boiled*	1590	260	110	(49)	270	1.1	(0.20)	2.2	2550	0.01	(23)	(30)
240	-, *weighed with shells*	600	99	43	(19)	100	0.4	(0.08)	0.8	970	Tr	(9)	(11)
241	Prawns, frozen, raw	780	92	83	34	140	1.0	0.14	1.1	1090	0.01	16	21
242	dried	N	N	240	N	1000	4.6	N	N	N	N	N	N
243	Scampi, in breadcrumbs, *frozen, fried in blended oil*	660	130	210	24	310	1.7	0.16	0.6	610	0.35	17	41
244	in breadcrumbs, frozen, *fried in sunflower oil*	660	130	210	24	310	1.7	0.16	0.6	610	0.35	17	41
245	Shrimps, *boiled*	3840	400	320	110	270	1.8	1.92	2.3	5850	0.03	(46)	(100)
246	-, *weighed with shells*	1270	130	110	36	89	0.6	0.63	0.8	1930	0.01	(15)	(33)
247	canned in brine, *drained*	980	100	110	49	150	5.1	0.23	1.9	1510	0.04	(21)	81
248	frozen	380	75	130	47	150	2.6	0.15	1.1	520	0.15	(43)	(100)
249	dried	4330	500	1200	290	710	21.3	0.96	3.7	6040	0.84	(170)	(400)

No. 16-	Food	Retinol µg	Carotene µg	Vitamin D µg	Vitamin E mg	Thiamin mg	Riboflavin mg	Niacin mg	Trypt 60 mg	Vitamin B6 mg	Vitamin B12 µg	Folate µg	Pantothenate mg	Biotin µg	Vitamin C mg
232	Crab, boiled	Tr	Tr	Tr	N	0.07	0.86	1.5	3.6	0.16	Tr	20	0.95	7	Tr
233	-, weighed with shell	Tr	Tr	Tr	N	0.02	0.30	0.5	1.3	0.06	Tr	7	0.33	3	Tr
234	canned in brine, drained	Tr	Tr	Tr	N	Tr	0.05	1.1	3.4	N	Tr	N	N	Tr	Tr
235	Crayfish, raw	Tr	Tr	Tr	N	0.05	0.04	1.9	2.8	0.07	2	30	0.57	N	Tr
236	Lobster, boiled	Tr	Tr	Tr	(1.47)	0.08	0.05	1.5	4.1	(0.08)	(3)	(9)	(1.00)	(7)	Tr
237	-, weighed with shell	Tr	Tr	Tr	(0.53)	0.03	0.02	0.5	1.5	(0.03)	(1)	(3)	(0.36)	(2)	Tr
238	Prawns, raw	Tr	Tr	Tr	(2.85)	0.04	(0.12)	(0.5)	3.8	(0.05)	(7)	N	(0.18)	(1)	Tr
239	boiled	Tr	Tr	Tr	N	(0.02)	(0.12)	(0.3)	4.8	(0.04)	(8)	N	(0.16)	(1)	Tr
240	-, weighed with shells	Tr	Tr	Tr	N	(0.01)	(0.05)	(0.1)	1.8	(0.02)	(3)	N	(0.06)	Tr	Tr
241	frozen, raw	Tr	Tr	Tr	(2.85)	0.02	0.12	0.5	0.1	0.05	7	N	0.18	1	Tr
242	dried	N	Tr	N	N	0.16	0.34	N	13.3	N	N	N	N	N	Tr
243	Scampi, in breadcrumbs, frozen, fried in blended oil	Tr	Tr	Tr	N	0.11	0.04	1.2	1.8	0.09	1	N	0.26	1	Tr
244	in breadcrumbs, frozen, fried in sunflower oil	Tr	Tr	Tr	N	0.11	0.04	1.2	1.8	0.09	1	N	0.26	1	Tr
245	Shrimps, boiled	N	Tr	Tr	N	0.01	0.01	1.0	4.4	0.08	(3)	(9)	0.24	1	Tr
246	-, weighed with shells	N	Tr	Tr	N	Tr	Tr	0.3	1.5	0.03	(1)	(3)	0.08	Tr	Tr
247	canned in brine, drained	N	Tr	Tr	N	0.01	0.02	0.8	3.9	0.03	2	15	0.35	1	Tr
248	frozen	2	Tr	Tr	N	Tr	0.02	0.5	3.1	(0.08)	3	14	(0.24)	(1)	Tr
249	dried	N	N	N	N	0.03	0.07	2.8	10.4	N	5	70	N	N	Tr

Molluscs

16-250 to 16-267

Composition of food per 100g

No. 16-	Food	Description and main data sources	Edible Proportion	Water g	Total Nitrogen g	Protein g	Fat g	Carbohydrate g	Energy value kcal	kJ
250	Abalone, canned in brine, drained	Ref 5	N	65.1	3.97	24.8	2.0	Tr	117	496
251	Clams, canned in brine, drained	Refs 5 and 1 and literature sources	N	69.9	2.56	16.0	0.6	1.9[a]	77	325
252	Cockles, boiled	11 samples from assorted outlets; fresh and frozen	1.00	83.0	1.92	12.0	0.6	Tr	53	226
253	bottled in vinegar, drained	10 samples, 5 brands	0.47	78.7	2.13	13.3	0.7	Tr	60	252
254	Cuttlefish, raw	Refs 5 and 1	0.79	80.8	2.58	16.1	0.7	0	71	300
255	Mussels, raw	Purchased alive	0.32	80.9	1.93	12.1	1.8	2.5[a]	74	312
256	boiled	11 fresh and frozen samples, boiled for 2 minutes	1.00	72.9	2.67	16.7	2.7	3.5[a]	104	440
257	-, weighed with shells	Calculated from boiled	0.27	19.7	0.72	4.5	0.7	0.9[a]	28	119
258	canned and bottled, drained	10 samples, in vinegar and brine	0.59	76.2	2.70	16.9	2.1	3.1[a]	98	415
259	Octopus, raw	Literature sources	0.79	82.1	2.86	17.9	1.3	Tr	83	352
260	Oysters, raw	Analytical and literature sources	1.00	85.7	1.72	10.8	1.3	2.7[a]	65	275
261	-, weighed with shells	Calculated from raw	0.14	12.0	0.24	1.5	0.2	0.4[a]	9	38
262	Scallops, steamed	Analytical and literature sources	1.00	73.1	3.71	23.2	1.4	3.4[a]	118	501
263	Squid, raw	Analytical and literature sources	N	80.5	2.46	15.4	1.7	(1.2)[a]	81	344
264	frozen, raw	5 samples, 4 brands	0.59	84.2	2.10	13.1	1.5	(1.0)[a]	70	294
265	in batter, fried in blended oil	Calculated from dissection of shop bought samples	1.00	62.4	1.85	11.5	10.0	15.7	195	815
266	-, fried in sunflower oil	Calculated from dissection of shop bought samples	1.00	62.4	1.85	11.5	10.0	15.7	195	815
267	dried	Ref 5 and calculation from raw	1.00	21.8	10.13	63.3	4.6	(4.8)[a]	313	1323

[a] As glycogen

No. 16-	Food	Starch g	Total sugars g	Dietary fibre Southgate method g	Englyst method g	Fatty acids Satd g	Mono-unsatd g	Poly-unsatd g	Cholesterol mg
250	**Abalone,** canned in brine, *drained*	Tr	Tr	0	0	N	N	N	N
251	**Clams,** canned in brine, *drained*	Tr	Tr	0	0	0.2	0.1	0.1	(67)
252	**Cockles,** *boiled*	Tr	Tr	0	0	0.2	0.1	0.2	53
253	bottled in vinegar, *drained*	Tr	Tr	0	0	0.2	0.1	0.2	(53)
254	**Cuttlefish,** raw	0	0	0	0	0.2	0.1	0.2	110
255	**Mussels,** raw	Tr	Tr	0	0	0.4	0.3	0.6	41
256	*boiled*	Tr	Tr	0	0	0.5	0.4	1.0	58
257	-, *weighed with shells*	Tr	Tr	0	0	0.1	0.1	0.3	16
258	canned and bottled, *drained*	Tr	Tr	0	0	0.4	0.3	0.7	(58)
259	**Octopus,** raw	Tr	Tr	0	0	0.3	0.2	0.5	48
260	**Oysters,** raw	Tr	Tr	0	0	(0.2)	(0.2)	(0.4)	(57)
261	-, *weighed with shells*	Tr	Tr	0	0	Tr	Tr	(0.1)	(8)
262	**Scallops,** *steamed*	Tr	Tr	0	0	0.4	0.1	0.4	47
263	**Squid,** raw	Tr	Tr	0	0	0.4	0.2	0.6	225
264	frozen, raw	Tr	Tr	0	0	0.3	0.2	0.5	200
265	in batter, *fried in blended oil*	12.9	2.2	0.6	0.5	2.1	3.3	3.7	145
266	-, *fried in sunflower oil*	12.9	2.2	0.6	0.5	2.2	2.3	4.7	145
267	dried	Tr	Tr	0	0	(1.0)	(0.6)	(1.7)	(615)

Inorganic constituents per 100g food

No. 16-	Food	mg										μg	
		Na	K	Ca	Mg	P	Fe	Cu	Zn	Cl	Mn	Se	I
250	Abalone, canned in brine, drained	990	110	10	N	150	8.8	N	N	N	N	N	N
251	Clams, canned in brine, drained	1200	(630)	74	(18)	280	8.0	(0.69)	(2.7)	970	0.03	N	N
252	Cockles, boiled	490	110	91	46	140	28.0	0.38	2.1	750	0.84	(43)	(160)
253	bottled in vinegar, drained	660	460	70	30	170	35.0	0.17	1.4	1060	0.54	43	160
254	Cuttlefish, raw	370	310	59	N	270	3.4	0.59	1.2	N	0.01	(65)	N
255	Mussels, raw	290	320	38	23	240	5.8	0.22	2.5	460	0.19	(51)	140
256	boiled	360	140	52	38	190	6.8	0.21	2.3	590	0.26	43	120
257	-, weighed with shells	96	38	14	10	51	1.8	0.06	0.6	160	0.07	12	31
258	canned and bottled, drained	460	59	49	28	180	13.0	0.25	2.7	790	0.47	(38)	(100)
259	Octopus, raw	N	230	33	N	170	1.2	0.40	1.7	N	0.02	(75)	20
260	Oysters, raw	510	260	140	42	210	5.7	7.50	59.2[a]	820	0.33	(23)	60
261	-, weighed with shells	71	36	19	6	30	0.8	1.05	8.3	110	0.05	(3)	8
262	Scallops, steamed	180	240	29	38	240	1.1	0.14	2.6	410	0.10	(51)	20
263	Squid, raw	110	280	13	28	190	0.5	0.98	1.1	N	0.02	(66)	20
264	frozen, raw	190	150	13	36	170	0.2	0.68	1.2	280	0.02	(66)	(20)
265	in batter, fried in blended oil	88	230	81	23	160	0.7	0.52	0.9	65	0.11	(35)	(19)
266	-, fried in sunflower oil	88	230	81	23	160	0.7	0.52	0.9	65	0.11	(35)	(18)
267	dried	940	610	55	(120)	620	4.0	(3.93)	(4.4)	N	(0.08)	N	N

[a] Levels ranged from 8mg to 90mg Zn per 100g

Molluscs

No. 16-	Food	Retinol µg	Carotene µg	Vitamin D µg	Vitamin E mg	Thiamin mg	Ribo-flavin mg	Niacin mg	Trypt 60 mg	Vitamin B6 mg	Vitamin B12 µg	Folate µg	Panto-thenate mg	Biotin µg	Vitamin C mg
250	Abalone, canned in brine, drained	N	30	N	N	0.01	0.04	2.6	5.3	N	N	N	N	N	Tr
251	Clams, canned in brine, drained	10	20	N	N	0.02	0.06	2.6	3.4	N	N	N	N	N	Tr
252	Cockles, boiled	40	Tr	Tr	N	0.05	0.11	1.2	2.6	0.04	47	N	0.27	9	Tr
253	bottled in vinegar, drained	51	Tr	Tr	N	(0.05)	(0.11)	(1.2)	2.8	(0.04)	(47)	N	(0.27)	(9)	Tr
254	Cuttlefish, raw	0	0	Tr	N	0.04	0.48	3.7	3.4	0.39	2	(13)	N	N	Tr
255	Mussels, raw	N	Tr	Tr	0.74	(0.02)	(0.35)	1.6	2.6	0.08	(19)	37	(0.39)	(7)	Tr
256	boiled	N	Tr	Tr	1.05	0.02	0.38	1.3	3.6	0.06	22	(37)	0.40	9	Tr
257	-, weighed with shells	N	Tr	Tr	0.28	0.01	0.10	0.3	1.0	0.02	6	(10)	0.11	2	Tr
258	canned and bottled, drained	N	Tr	Tr	(1.05)	(0.02)	(0.38)	(1.3)	3.6	(0.06)	(22)	(37)	(0.40)	(9)	Tr
259	Octopus, raw	5	0	Tr	N	0.07	0.07	5.0	3.8	0.36	N	N	N	N	0
260	Oysters, raw	75	Tr	1.0	0.85	0.15	0.19	1.8	2.3	0.16	17	N	0.37	10	Tr
261	-, weighed with shells	11	Tr	0.1	0.12	0.02	0.03	0.3	0.3	0.02	2	N	0.05	1	Tr
262	Scallops, steamed	Tr	Tr	Tr	N	0.08	0.05	0.9	5.0	N	9	18	0.16	2	Tr
263	Squid, raw	15	0	Tr	1.20	0.10	0.12	3.4	3.3	0.69	3	13	0.68	N	0
264	frozen, raw	(15)	0	Tr	(1.20)	0.05	0.02	2.1	2.8	(0.69)	3	2	(0.68)	N	0
265	in batter, fried in blended oil	46	13	0.1	2.35	0.10	0.13	1.6	2.5	0.32	2	15	0.52	N	Tr
266	-, fried in sunflower oil	46	13	0.1	4.00	0.10	0.13	1.6	2.5	0.32	2	15	0.52	N	Tr
267	dried	N	0	N	N	0.13	0.17	6.8	13.5	N	N	N	N	N	0

No. Food 16-	Description and main data sources	Edible Proportion	Water g	Total Nitrogen g	Protein g	Fat g	Carbo- hydrate g	Energy value kcal	kJ
268 **Whelks**, *boiled*	10 samples from assorted outlets, boiled in salted water	1.00	73.9	3.12	19.5	1.2	Tr[a]	89	376
269 -, *weighed with shells*	Calculated from boiled	0.34	25.1	1.06	6.6	0.4	Tr[a]	30	128
270 **Winkles**, *boiled*	11 samples, from stalls and fishmongers. Purchased cooked	1.00	73.0	2.46	15.4	1.2	Tr[a]	72	306
271 -, *weighed with shells*	Calculated from boiled	0.19	13.9	0.47	2.9	0.2	Tr[a]	14	58

[a] As glycogen

82

No. Food 16-	Starch g	Total sugars g	Dietary fibre Southgate method g	Dietary fibre Englyst method g	Fatty acids Satd g	Fatty acids Mono-unsatd g	Fatty acids Poly-unsatd g	Cholesterol mg
268 **Whelks**, *boiled*	Tr	Tr	0	0	0.2	0.2	0.3	125
269 *-, weighed with shells*	Tr	Tr	0	0	0.1	0.1	0.1	43
270 **Winkles**, *boiled*	Tr	Tr	0	0	0.2	0.2	0.4	105
271 *-, weighed with shells*	Tr	Tr	0	0	Tr	Tr	0.1	20

Molluscs

16-268 to 16-271

Inorganic constituents per 100g food

No. 16-	Food	Na	K	Ca	Mg	P	Fe	Cu	Zn	Cl	Mn	Se	I
							mg					μg	
268	Whelks, *boiled*	280	190	84	87	140	3.3	6.59[a]	12.1[b]	480	0.12	N	N
269	-, *weighed with shells*	95	65	29	30	48	1.1	2.24	4.1	160	0.04	N	N
270	Winkles, *boiled*	750	220	130	310	160	10.2	1.70	3.3	1160	1.20	N	80
271	-, *weighed with shells*	140	42	25	58	29	1.9	0.32	0.6	220	0.23	N	15

[a] Levels ranged from 1.2mg to 18.5mg Cu per 100g

[b] Levels ranged from 5.6mg to 20.0mg Zn per 100g

Molluscs

No. Food 16-	Retinol µg	Carotene µg	Vitamin D µg	Vitamin E mg	Thiamin mg	Ribo-flavin mg	Niacin mg	Trypt 60 mg	Vitamin B6 mg	Vitamin B12 µg	Folate µg	Panto-thenate mg	Biotin µg	Vitamin C mg
268 **Whelks**, *boiled*	N	Tr	Tr	0.80	0.04	0.17	1.3	4.2	0.09	21	(6)	0.55	6	Tr
269 *-, weighed with shells*	N	Tr	Tr	0.27	0.01	0.06	0.4	1.4	0.03	7	(2)	0.19	2	Tr
270 **Winkles**, *boiled*	N	Tr	Tr	3.90	0.29	0.38	1.7	3.3	0.10	36	N	0.38	3	Tr
271 *-, weighed with shells*	N	Tr	Tr	0.74	0.06	0.07	0.3	0.6	0.02	7	N	0.07	1	Tr

Fish products and dishes

Composition of food per 100g

No. 16-	Food	Description and main data sources	Edible Proportion	Water g	Total Nitrogen g	Protein g	Fat g	Carbo-hydrate g	Energy value kcal	kJ
272	Caviare, bottled in brine, *drained*	Ref 3 and literature sources. Lumpfish roe	0.94	74.3	1.98[a]	10.9[b]	5.4	0	92	385
273	Crabsticks	10 samples from assorted outlets. Crab flavoured minced fish sticks	1.00	76.6	1.60	10.0	0.4	6.6	68	290
274	Curry, fish, Bangladeshi	Recipe	1.00	70.7	1.96	12.2	7.9	1.5	124	517
275	fish and vegetable, Bangladeshi	Recipe	1.00	76.6	1.46	9.1	8.4	1.4	117	487
276	haddock, Bengali	Recipe	1.00	57.4	1.94	12.1	22.0	4.1	262	1084
277	herring, Bengali	Recipe	1.00	51.2	1.83	11.4	29.4	4.2	326	1348
278	prawn and mushroom	Recipe	1.00	66.7	1.36	7.3	14.4	2.5	168	695
279	Fish balls, *steamed*	7 varieties from different Chinese shops	1.00	82.2	1.89	11.8	0.5	5.5	74	313
280	Fish cakes, frozen	11 packets, 7 brands. Coated in breadcrumbs	1.00	64.9	1.38	8.6	3.9	16.7	132	558
281	*grilled*	Samples as frozen; grilled 5 minutes each side	1.00	58.6	1.58	9.9	4.5	19.7	154	650
282	*fried in blended oil*	Samples as frozen; shallow fried 5 minutes each side	1.00	55.2	1.38	8.6	13.4	16.8	218	911
283	*fried in lard*	Samples as frozen; shallow fried 5 minutes each side	1.00	55.2	1.38	8.6	13.4	16.8	218	911
284	*fried in sunflower oil*	Samples as frozen; shallow fried 5 minutes each side	1.00	55.2	1.38	8.6	13.4	16.8	218	911
285	*cod, homemade*	Recipe	1.00	58.2	1.51	9.3	16.6	14.4	241	1003
286	*salmon, homemade*	Recipe	1.00	54.2	1.69	10.4	19.7	14.4	273	1137
287	Fish fingers, cod, frozen	11 packets, 8 brands. Coated in breadcrumbs	1.00	63.1	1.86	11.6	7.8	14.2	170	713
288	*grilled*	Samples as frozen; grilled 5 minutes each side	1.00	55.7	2.29	14.3	8.9	16.6	200	838
289	*fried in blended oil*	Samples as frozen; shallow fried 4 minutes each side	1.00	53.8	2.11	13.2	14.1	15.5	238	994

[a] Includes 0.24g purine nitrogen per 100g

[b] (Total N - non-protein N) x 6.25

Fish products and dishes

Composition of food per 100g

No. 16-	Food	Starch g	Total sugars g	Dietary fibre Southgate method g	Dietary fibre Englyst method g	Fatty acids Satd g	Fatty acids Mono- unsatd g	Fatty acids Poly- unsatd g	Cholest- erol mg
272	**Caviare, bottled in brine,** *drained*	0	0	0	0	0.8	1.2	1.8	285
273	**Crabsticks**	6.6	Tr	0	0	N	N	N	(39)
274	**Curry,** fish, Bangladeshi	Tr	1.0	0.3	0.3	N	N	N	N
275	fish and vegetable, Bangladeshi	0.1	1.1	0.5	0.5	N	N	N	N
276	haddock, Bengali	Tr	3.0	0.9	0.8	14.2	5.3	0.9	81
277	herring, Bengali	0.1	3.0	0.9	0.8	16.1	8.5	2.4	90
278	prawn and mushroom	0.4	1.6	1.5	1.0	1.5	5.0	6.9	(71)
279	**Fish balls,** *steamed*	5.5	Tr	0	0	N	N	N	N
280	**Fish cakes,** frozen	16.7	Tr	0.7	N	0.5	1.3	0.9	21
281	*grilled*	19.7	Tr	0.8	N	0.6	1.5	1.0	25
282	*fried in blended oil*	16.8	Tr	0.7	N	1.8	5.4	5.6	21
283	*fried in lard*	16.8	Tr	0.7	N	4.4	6.1	2.3	(29)
284	*fried in sunflower oil*	16.8	Tr	0.7	N	1.9	4.0	6.8	21
285	cod, homemade	13.8	0.6	1.0	0.7	2.4	6.0	6.9	75
286	salmon, homemade	13.8	0.6	1.0	0.7	2.9	7.3	7.8	78
287	**Fish fingers,** cod, frozen	14.2	Tr	0.6	0.6	2.5	3.0	2.0	29[a]
288	*grilled*	16.6	Tr	0.7	0.7	2.8	3.4	2.3	35
289	*fried in blended oil*	15.5	Tr	0.6	0.6	(3.6)	(5.3)	(4.6)	32

[a] Fish fingers coated in batter contain 34mg cholesterol per 100g

87

Fish products and dishes

Inorganic constituents per 100g food

No. 16-	Food	mg										µg	
		Na	K	Ca	Mg	P	Fe	Cu	Zn	Cl	Mn	Se	I
272	Caviare, bottled in brine,												
	drained	2120	60	10	3	N	0.5	N	1.1	N	N	N	N
273	Crabsticks	700	58	13	7	47	0.8	0.28	3.7	1010	0.08	N	N
274	Curry, fish, Bangladeshi	N	240	N	N	260	0.8	0.08	0.3	N	N	N	N
275	fish and vegetable, Bangladeshi	N	210	N	N	150	0.9	0.06	0.2	N	N	N	N
276	haddock, Bengali	280	360	27	22	140	0.5	0.05	0.4	430	0.09	16	160
277	herring, Bengali	310	340	54	27	160	1.2	0.12	0.7	480	0.11	21	29
278	prawn and mushroom	440	320	41	(22)	120	0.8	(0.45)	0.9	740	0.10	11	12
279	Fish balls, steamed	750	110	26	13	100	0.2	0.07	0.3	1030	0.02	N	N
280	Fish cakes, frozen	640	290	140	24	160	1.4	0.07	0.5	800	0.26	N	N
281	grilled	600	280	150	22	140	1.1	0.04	0.5	870	0.26	N	N
282	fried in blended oil	510	230	110	18	110	0.8	0.03	0.4	(640)	0.20	N	N
283	fried in lard	510	230	110	18	110	0.8	0.03	0.4	(640)	0.20	N	N
284	fried in sunflower oil	510	230	110	18	110	0.8	0.03	0.4	(640)	0.20	N	N
285	cod, homemade	150	210	44	18	120	0.8	0.06	0.5	230	0.11	14	23
286	salmon, homemade	130	240	36	20	140	0.9	0.06	0.6	200	0.11	(11)	24
287	Fish fingers, cod, frozen	470	300	88	28	240	0.7	0.04	0.3	500	0.30	(19)	110[a]
288	grilled	440	290	92	22	220	0.8	0.02	0.4	590	0.21	(23)	110
289	fried in blended oil	450	260	85	22	220	0.8	0.04	0.3	550	0.19	(21)	(120)

[a] Levels ranged from 72mg to 230µg I per 100g

Fish products and dishes

No. 16-	Food	Retinol µg	Carotene µg	Vitamin D µg	Vitamin E mg	Thiamin mg	Ribo- flavin mg	Niacin mg	Trypt 60 mg	Vitamin B6 mg	Vitamin B12 µg	Folate µg	Panto- thenate mg	Biotin µg	Vitamin C mg
272	Caviare, bottled in brine, drained	Tr	Tr	N	N	Tr	0.09	Tr	2.0	N	N	N	N	N	0
273	Crabsticks	Tr	Tr	Tr	N	0.01	0.06	0.2	1.9	0.02	1	N	N	N	Tr
274	Curry, fish, Bangladeshi	N	12	N	(1.56)	(0.04)	(0.02)	0.6	2.3	N	4	N	N	N	1
275	fish and vegetable, Bangladeshi	N	120	N	(0.98)	(0.05)	(0.04)	1.4	1.7	N	3	N	N	N	4
276	haddock, Bengali	215	285	0.2	1.19	0.07	0.04	2.5	2.3	0.31	1	13	0.22	2	7
277	herring, Bengali	240	285	11.4	1.41	0.06	0.13	2.4	2.1	0.34	7	13	0.48	4	7
278	prawn and mushroom	Tr	43	Tr	3.59	0.07	0.16	1.6	1.5	0.13	2	28	0.90	6	4
279	Fish balls, steamed	N	Tr	N	N	0.02	0.03	0.8	2.2	N	1	4	N	N	0
280	Fish cakes, frozen	Tr	Tr	Tr	N	0.09	0.14	1.7	1.6	0.28	1	N	0.33	1	Tr
281	grilled	Tr	Tr	Tr	N	0.08	0.16	1.9	1.8	0.29	1	N	0.39	1	Tr
282	fried in blended oil	Tr	Tr	Tr	N	0.07	0.14	1.6	1.6	0.25	1	N	0.33	1	Tr
283	fried in lard	Tr	Tr	N	N	0.07	0.14	1.6	1.6	0.25	1	N	0.33	1	Tr
284	fried in sunflower oil	Tr	Tr	Tr	N	0.07	0.14	1.6	1.6	0.25	1	N	0.33	1	Tr
285	cod, homemade	44	12	0.4	3.44	0.13	0.12	1.1	2.0	0.16	1	22	0.56	17	1
286	salmon, homemade	48	12	3.0	4.05	0.13	0.10	2.2	2.2	0.30	2	24	0.61	5	1
287	Fish fingers, cod, frozen	Tr	Tr	Tr	N	0.11	0.08	1.5	2.2	0.14	1	15	0.30	1	Tr
288	grilled	Tr	Tr	Tr	N	0.12	0.08	1.7	2.7	0.17	1	16	0.35	1	Tr
289	fried in blended oil	Tr	Tr	Tr	N	0.11	0.07	1.6	2.5	0.15	1	16	0.32	1	Tr

Fish products and dishes

Composition of food per 100g

No. 16-	Food	Description and main data sources	Edible Proportion	Water g	Total Nitrogen g	Protein g	Fat g	Carbo-hydrate g	Energy value kcal	Energy value kJ
290	Fish fingers, cod, fried in lard	Samples as frozen; shallow fried 4 minutes each side	1.00	53.8	2.11	13.2	14.1	15.5	238	994
291	-, fried in sunflower oil	Samples as frozen; shallow fried 4 minutes each side	1.00	53.8	2.11	13.2	14.1	15.5	238	994
292	economy, frozen	10 packets, 8 brands. Coated in breadcrumbs	1.00	62.1	1.84	11.5	8.2	13.7	171	718
293	Fish paste	30 samples, sardine, crab, lobster and salmon	1.00	67.1	2.45	15.3	10.5	3.7	170	708
294	Fish pie	Recipe	1.00	74.9	1.07	6.7	5.4	11.5	119	498
295	Fisherman's pie, retail	Calculated from manufacturers proportions	1.00	75.3	1.42	8.9	5.4	8.9	118	493
296	Kedgeree	Recipe	1.00	64.8	2.56	15.9	8.7	7.8	171	715
297	Mackerel pâté, smoked	6 samples, 5 brands	1.00	48.4	2.14	13.4	34.4	1.3	368	1521
298	Pilau, prawn	Recipe	1.00	64.3	0.90	5.5	3.3	24.9	145	612
299	Roe, cod, hard, raw	6 samples from assorted outlets	1.00	73.7	3.47	21.7	1.9	0	104	439
300	-, fried in blended oil	Parboiled, slices, coated in crumbs and fried in blended oil. Some nutrients calculated from raw	1.00	62.0	3.34	20.9	11.9	3.0	202	844
301	coated in batter, fried	7 samples from fish and chip shops	1.00	64.4	1.98	12.4	11.8	8.9	189	790
302	herring, soft, raw	10 samples from assorted outlets	1.00	80.0	2.98[a]	16.8[b]	2.6	0	91	382
303	-, fried in blended oil	Rolled in flour and fried in blended oil. Some nutrients calculated from raw	1.00	52.3	4.66	26.3	15.8	4.7	265	1107
304	Salmon en croûte, retail	Calculated from manufacturers' proportions	1.00	49.3	1.93	11.8	19.1	18.0	288	1202
305	Seafood pasta, retail	Calculated from manufacturers' proportions	1.00	77.1	1.45	8.9	4.8	7.6	110	460

[a] Includes 0.24g purine nitrogen per 100g

[b] (Total N- non-protein N) x 6.25

Fish products and dishes

Composition of food per 100g

No. 16-	Food	Starch g	Total sugars g	Dietary fibre Southgate method g	Englyst method g	Fatty acids Satd g	Mono-unsatd g	Poly-unsatd g	Cholesterol mg
290	**Fish fingers**, cod, fried in lard	15.5	Tr	0.6	0.6	(4.9)	(5.7)	(3.0)	(37)
291	-, fried in sunflower oil	15.5	Tr	0.6	0.6	(3.6)	(4.6)	(5.2)	32
292	economy, frozen	13.7	Tr	(0.6)	(0.6)	(2.6)	(3.1)	(2.1)	(29)
293	**Fish paste**	3.2	0.5	0.2	(0.2)	N	N	N	N
294	**Fish pie**	9.9	1.6	0.8	0.7	1.9	1.9	1.2	18
295	**Fisherman's pie**, retail	7.2	1.7	0.6	0.5	1.8	1.3	0.7	N
296	**Kedgeree**	7.7	0.1	0.2	0.1	(2.5)	(3.3)	(1.8)	(135)
297	**Mackerel pâté**, smoked	0.6	0.7	Tr	Tr	N	N	N	N
298	**Pilau**, prawn	23.8	0.7	0.8	0.3	1.9	0.7	0.1	(51)
299	**Roe**, cod, hard, raw	0	0	0	0	0.4	0.4	0.6	330
300	-, *fried in blended oil*	3.0	Tr	0.1	0.1	1.6	4.1	5.7	315
301	coated in batter, *fried*	8.9	Tr	0.3	0.2	N	N	N	N
302	herring, soft, raw	0	0	0	0	0.5	0.6	0.7	575
303	-, *fried in blended oil*	4.7	Tr	0.2	0.2	(2.6)	(5.4)	(7.1)	895
304	**Salmon en croûte**, retail	17.1	0.9	1.1	N	3.1	2.9	1.4	31
305	**Seafood pasta**, retail	5.9	1.6	0.6	0.4	2.8	1.2	0.3	41

No. 16-	Food	mg										μg	
		Na	K	Ca	Mg	P	Fe	Cu	Zn	Cl	Mn	Se	I
290	Fish fingers, cod, fried in lard	450	260	85	22	220	0.8	0.04	0.3	550	0.19	(21)	(120)
291	-, fried in sunflower oil	450	260	85	22	220	0.8	0.04	0.3	550	0.19	(21)	(120)
292	economy, frozen	340	460	140	22	270	0.9	0.04	0.3	540	0.23	19	78
293	Fish paste	600	300	280	33	310	9.0^a	0.60	2.0	940	N	N	310^b
294	Fish pie	100	260	47	17	93	0.4	0.04	0.4	170	0.07	11	(15)
295	Fisherman's pie, retail	130	260	71	21	120	0.4	0.04	0.5	(200)	0.05	13	N
296	Kedgeree	630	270	31	22	180	0.6	0.06	0.7	980	0.08	23	(200)
297	Mackerel pâté, smoked	730	230	28	21	160	1.0	0.10	0.7	1180	0.02	26	N
298	Pilau, prawn	290	97	24	(13)	75	0.4	(0.09)	0.7	460	0.27	6	10
299	Roe, cod, hard, raw	110	170	11	9	320	1.0	0.24	3.5	220	0.12	N	N
300	-, fried in blended oil	120	170	13	9	300	1.0	0.22	3.3	230	0.12	N	N
301	coated in batter, fried	510	270	67	27	210	1.4	0.13	1.5	790	0.44	25	N
302	herring, soft, raw	120	200	5	5	620	0.5	0.07	0.8	180	0.01	N	N
303	-, fried in blended oil	180	320	16	9	970	0.9	0.11	1.2	290	0.06	N	N
304	Salmon en croûte, retail	190	200	47	18	140	0.6	0.08	0.5	320	0.13	N	N
305	Seafood pasta, retail	170	190	38	20	100	0.4	0.07	0.5	260	0.11	12	39

^{a}Iron oxides are often added as a colourant

^{b}Crab paste contains 240μg I and salmon paste 370μg I per 100g

Fish products and dishes

No. 16-	Food	Retinol µg	Carotene µg	Vitamin D µg	Vitamin E mg	Thiamin mg	Ribo-flavin mg	Niacin mg	Trypt 60 mg	Vitamin B6 mg	Vitamin B12 µg	Folate µg	Panto-thenate mg	Biotin µg	Vitamin C mg
290	Fish fingers, cod, fried in lard	Tr	Tr	N	N	0.11	0.07	1.6	2.5	0.15	1	16	0.32	1	Tr
291	-, fried in sunflower oil	Tr	Tr	Tr	N	0.11	0.07	1.6	2.5	0.15	1	16	0.32	1	Tr
292	economy, frozen	Tr	Tr	Tr	N	0.10	0.08	1.2	2.2	0.12	1	(15)	0.28	2	Tr
293	Fish paste	19[a]	Tr	N	0.87	0.02	0.20	4.1	2.9	N	N	N	N	N	Tr
294	Fish pie	53	44	0.4	0.45	0.11	0.08	1.0	1.3	0.22	0	14	0.41	13	3
295	Fisherman's pie, retail	57	32	0.2	(0.33)	0.06	0.07	0.9	1.8	(0.14)	(0)	(10)	(0.22)	(1)	2
296	Kedgeree	100	49	1.0	N	0.05	0.21	2.1	3.4	0.28	2	N	0.52	6	Tr
297	Mackerel pâté, smoked	165	N	3.3	1.11	0.11	0.44	4.9	2.5	0.28	18	3	0.90	6	N
298	Pilau, prawn	28	17	Tr	0.13	0.04	0.03	0.5	1.2	0.12	1	8	(0.16)	(1)	1
299	Roe, cod, hard, raw	79	0	18.0	6.20	0.78	0.49	1.0	4.1	0.32	10	N	3.00	13	Tr
300	-, fried in blended oil	75	Tr	17.0	N	0.59	0.37	1.0	3.9	0.28	11	N	2.60	15	Tr
301	coated in batter, fried	N	N	N	N	0.09	0.22	0.6	2.3	0.08	N	22	1.57	8	Tr
302	herring, soft, raw	N	0	4.0	2.80	Tr	0.29	1.3	3.1	N	5	N	0.49	N	Tr
303	-, fried in blended oil	N	Tr	6.3	N	Tr	0.36	1.7	4.9	N	6	N	0.61	N	Tr
304	Salmon en croûte, retail	30	13	(3.4)	N	0.13	0.07	2.5	2.3	0.31	2	12	N	N	Tr
305	Seafood pasta, retail	52	52	0	0.40	0.03	0.06	1.0	1.7	0.09	1	6	0.21	1	1

[a] Salmon paste contains 49µg retinol per 100g

Fish products and dishes

No. Food 16-	Description and main data sources	Edible Proportion	Water g	Total Nitrogen g	Protein g	Fat g	Carbo-hydrate g	Energy value kcal	kJ
306 **Seafood cocktail**	Recipe. Mussels, crabsticks, prawns, squid and cockles	1.00	75.3	2.50	15.6	1.5	2.9	87	369
307 **Taramasalata**	10 assorted samples. Greek dish based on cod's roe	1.00	35.9	0.51	3.2	52.9	4.1	504	2077
308 **Tuna pâté**	Calculated from manufacturers' proportions	1.00	61.8	2.71	17.0	18.6	0.4	236	982

94

Fish products and dishes

| No. Food | Starch | Total sugars | Dietary fibre Southgate method | Dietary fibre Englyst method | Fatty acids Satd | Fatty acids Mono-unsatd | Fatty acids Poly-unsatd | Cholest-erol |
16-	g	g	g	g	g	g	g	mg
306 **Seafood cocktail**	1.4	Tr	0	0	0.3	0.2	0.5	(115)
307 **Taramasalata**	4.1	Tr	Tr	Tr	4.1	29.3	16.7	25
308 **Tuna pâté**	Tr	0.3	Tr	Tr	7.8	4.4	5.3	72

95

Fish products and dishes

Inorganic constituents per 100g food

No. Food	Na	K	Ca	Mg	P	Fe	Cu	Zn	Cl	Mn	Se	I
16-						mg					μg	
306 Seafood cocktail	620	160	52	(32)	170	5.6	(0.36)	2.4	940	0.20	N	N
307 Taramasalata	650	60	21	6	50	0.4	N	0.4	1040	0.12	N	N
308 Tuna pâté	390	170	12	21	130	0.8	0.05	0.5	640	Tr	N	17

Fish products and dishes

No. Food 16-	Retinol	Carotene	Vitamin D	Vitamin E	Thiamin	Ribo-flavin	Niacin	Trypt 60	Vitamin B6	Vitamin B12	Folate	Panto-thenate	Biotin	Vitamin C
	µg	µg	µg	mg	mg	mg	mg	mg	mg	µg	µg	mg	µg	mg
306 Seafood cocktail	6	Tr	Tr	(0.56)	0.03	0.20	1.2	3.3	0.14	15	N	(0.30)	N	Tr
307 Taramasalata	N	N	N	N	0.08	0.10	0.3	0.6	N	3	4	N	N	1
308 Tuna pâté	N	59	2.9	2.71	0.02	0.09	10.1	3.2	0.34	3	4	0.22	1	1

Appendices

INDIVIDUAL FATTY ACIDS

Values for individual fatty acids in fish and fish products are given for foods where total fat is more than or equal to 0.5 gram per 100 g. The values are in grams per 100g food. Values included below are those obtained by analysis only, and do not include values derived by calculation or estimation.

Names of the fatty acids occurring in the tables

Carbon: Double bonds	Systematic name	Common name
Saturated acids		
10:0	Decanoic acid	Capric acid
12:0	Dodecanoic acid	Lauric acid
14:0	Tetradecanoic acid	Myristic acid
16:0	Hexadecanoic acid	Palmitic acid
17:0	Heptadecanoic acid	Margaric acid
18:0	Octadecanoic acid	Stearic acid
20:0	Eicosanoic acid	Arachidic acid Arachic acid
22:0	Docosanoic acid	Behenic acid
Monounsaturated acids		
10:1	Decenoic acid	
14:1	Tetradecenoic acid	Myristoleic acid
15:1	Pentadecenoic acid	
16:1	cis-9-Hexadecenoic acid	Palmitoleic acid
18:1	cis-9-Octadecenoic acid	Oleic acid
20:1	Eicosenoic acid	Eicosenic acid Gadoleic acid
22:1	Docosenoic acid	
Polyunsaturated acids		
16:2	Hexadecadienoic acid	
16:4	Hexadecatetraenoic acid	
16:UNID	Unidentified C16 fatty acids	
18:2	Octadecadienoic acid	Linoleic acid
18:3[a]	Octadecatrienoic acid	Linolenic acid
18:4	Octadecatetraenoic acid	Stearidonic acid
20:3	Eicosatrienoic acid	
20:4[a]	Eicosatetraenoic acid	Arachidonic acid
20:5[a]	Eicosapentaenoic acid	
21:5[a]	Heneicosapentaenoic acid	
22:2	Docosadienoic acid	
22:4	Docosatetraenoic acid	
22:5[a]	Docosapentaenoic acid	Clupanodonic acid
22:6[a]	Docosahexaenoic acid	Cervonic acid
22:UNID	Unidentified C22 fatty acids	

[a] These are the main n-3 (ω-3) fatty acids in fatty fish, 18:3 and 22:5 contain both n-3 (ω-3) and n-6 (ω-6) isomers

Fish and fish products

Fatty Acids per 100g food

No. 16-	Food	Saturated						Monounsaturated					
		12:0	14:0	15:0	16:0	17:0	18:0	14:1	16:1	17:1	18:1	20:1	22:1
	White fish												
2	**Bass, Sea**, raw	0	Tr	Tr	0.3	0	0.1	0	0.1	0	0.4	0.1	0
7	**Catfish**, raw	0	0	0	0	0	0	0	0	0	0	0	0
8	*steamed*	0	0	0	0	0	0	0	0	0	0	0	0
11	**Chital**, raw	0	0	0	0	0	0	0	0	0	0	0	0
12	**Cod**, raw	0	Tr	Tr	0.1	Tr	Tr	0	Tr	0	0.1	Tr	Tr
17	*steamed*	0	Tr	Tr	0.1	Tr	Tr	0	Tr	0	0.1	Tr	Tr
19	*frozen, raw*	0	Tr	Tr	0.1	Tr	Tr	0	Tr	0	Tr	Tr	Tr
25	*coated in batter, frozen, baked* [a]	0.1	0.1	Tr	2.3	Tr	1.0	0	Tr	0	5.7	0.1	0.1
28	*smoked, raw*	0	Tr	Tr	0.1	Tr	Tr	0	Tr	0	Tr	Tr	Tr
31	**Coley**, raw	0	Tr	Tr	0.1	Tr	Tr	0	Tr	Tr	0.1	Tr	Tr
32	*steamed*	0	Tr	Tr	0.1	Tr	Tr	0	Tr	Tr	0.2	Tr	0.1
34	*frozen, raw*	0	Tr	Tr	0.1	Tr	Tr	0	Tr	Tr	0.1	Tr	Tr
35	**Conger eel**, raw	0	0	0	0	0	0	0	0	0	0	0	0
36	*grilled*	0	0.1	Tr	0.5	Tr	0.1	0	0.3	0.1	0.5	0.2	0.1
37	*-, weighed with bones and skin*	0	0.1	Tr	0.3	Tr	0.1	0	0.2	Tr	0.3	0.1	Tr
38	**Dab**, raw	0	0	0	0	0	0	0	0	0	0	0	0
39	**Dover sole**, raw	0	0	0	0	0	0	0	0	0	0	0	0
43	**Flying fish**, raw	0	0	0	0	0	0	0	0	0	0	0	0
44	**Haddock**, raw	0	Tr	Tr	0.1	Tr	Tr	0	Tr	0	0.1	Tr	Tr
49	*steamed*	0	Tr	Tr	0.1	Tr	Tr	0	Tr	0	0.1	Tr	Tr

[a] Cod, coated in batter contains 0.1g 20:0 per 100g

Fish and fish products

Fatty Acids per 100g food

No.	Food	Polyunsaturated											
		16:4	16:UNID	18:2	18:3	18:4	20:3	20:4	20:5	21:5	22:4	22:5	22:6
16-													
White fish													
2	**Bass, Sea**, raw	0	0	Tr	0	0	0	Tr	0.1	0	0	Tr	0.4
7	**Catfish**, raw	0	0	0	0	0	0	0	0	0	0	0	0
8	*steamed*	0	0	0	0	0	0	0	0	0	0	0	0
11	**Chital**, raw	0	0	0	0	0	0	0	0	0	0	0	0
12	**Cod**, raw	0	0	Tr	Tr	Tr	0	Tr	0.1	0	0	Tr	0.2
17	*steamed*	0	0	Tr	Tr	Tr	0	Tr	0.1	0	0	Tr	0.2
19	frozen, raw	0	0	Tr	Tr	Tr	0	Tr	0.1	0	0	Tr	0.1
25	coated in batter, frozen, *baked*	0	0	1.5	0.1	0	0	0	0	0	0	Tr	0.1
28	smoked, raw	0	0	Tr	Tr	Tr	0	Tr	0.1	0	0	Tr	0.1
31	**Coley**, raw	0	0	Tr	Tr	Tr	0	0.1	0	0	0	Tr	0.2
32	*steamed*	0	0	Tr	Tr	Tr	0	0.1	0	0	0	Tr	0.2
34	frozen, raw	0	0	Tr	Tr	Tr	0	0.1	0	0	0	Tr	0.2
35	**Conger eel**, raw	0	0	0	0	0	0	0	0	0	0	0	0
36	*grilled*	0	0	Tr	Tr	0.1	0	1.2	0	0	0	0.1	0.4
37	-, *weighed with bones and skin*	0	0	Tr	Tr	Tr	0	0.6	0	0	0	0.1	0.2
38	**Dab**, raw	0	0	0	0	0	0	0	0	0	0	0	0
39	**Dover sole**, raw	0	0	0	0	0	0	0	0	0	0	0	0
43	**Flying fish**, raw	0	0	0	0	0	0	0	0	0	0	0	0
44	**Haddock**, raw	0	0	Tr	Tr	Tr	0	Tr	0.1	0	Tr	Tr	0.1
49	*steamed*	0	0	Tr	Tr	Tr	0	Tr	0.1	0	0	Tr	0.1

Fish and fish products

Fatty Acids per 100g food

No. 16-	Food	Saturated						Monounsaturated					
		12:0	14:0	15:0	16:0	17:0	18:0	14:1	16:1	17:1	18:1	20:1	22:1
80	**John Dory**, raw	0	Tr	0	0.2	0	0.1	Tr	Tr	0	0.1	Tr	Tr
85	**Lemon sole**, *steamed*	0	Tr	Tr	0.1	Tr	Tr	0	0.1	Tr	0.1	Tr	0
92	**Monkfish**, raw	0	Tr	0	Tr	0	Tr	0	Tr	0	0.1	0	0
102	**Plaice**, raw	0	Tr	Tr	0.2	Tr	Tr	0	0.1	Tr	0.2	Tr	Tr
103	*grilled*	0	Tr	Tr	0.2	Tr	Tr	0	0.2	Tr	0.2	Tr	Tr
105	frozen, raw	0	Tr	Tr	0.1	Tr	Tr	0	0.1	Tr	0.2	Tr	Tr
106	-, *grilled*	0	Tr	Tr	0.2	Tr	Tr	0	0.2	Tr	0.3	0.1	Tr
108	-, *steamed*	0	Tr	Tr	0.2	Tr	Tr	0	0.2	Tr	0.2	Tr	Tr
133	**Rock Salmon/Dogfish**, raw	0	0.1	Tr	1.0	Tr	0.2	0	0.4	0.1	1.4	0.5	0.3
142	**Shark**, raw	0	Tr	Tr	0.1	0	0.1	0	Tr	0	0.1	Tr	Tr
143	**Skate**, raw	0	Tr	Tr	Tr	Tr	Tr	0	Tr	Tr	Tr	0	0
159	**Whiting**, raw	0	Tr	Tr	0.1	Tr	Tr	0	Tr	Tr	0.1	Tr	Tr
160	*steamed*	0	Tr	Tr	0.1	Tr	Tr	0	Tr	Tr	0.1	0	Tr
	Fatty fish												
174	**Eel**, jellied	Tr	0.3	Tr	1.2	0.1	0.3	0.1	0.9	Tr	2.4	0.1	0
175	**Herring**, raw[a]	Tr	0.9	0.1	1.9	Tr	0.3	Tr	1.1	0.1	1.8	1.0	1.5
176	*grilled*[a]	Tr	0.8	0.1	1.6	Tr	0.2	Tr	0.9	0.1	1.5	0.9	1.3
187	**Kipper**, raw[b]	Tr	0.9	0.1	1.6	0.1	0.2	Tr	0.9	0.1	1.9	2.4	3.8
188	*grilled*[b]	Tr	0.9	0.1	1.7	0.1	0.2	0.1	1.0	0.1	2.1	2.6	4.2
191	**Mackerel**, raw	0	1.0	0.1	1.9	0.1	0.3	0	0.6	Tr	1.9	2.1	3.2
194	*grilled*	0	1.1	0.1	2.0	0.1	0.3	0	0.6	Tr	2.1	2.3	3.4

[a] Herring contains 0.1g 15:1 per 100g

[b] Kipper contains 0.1g 10:1 per 100g

Fish and fish products

No. 16-	Food	Polyunsaturated											
		16:4	16:UNID	18:2	18:3	18:4	20:3	20:4	20:5	21:5	22:4	22:5	22:6
80	John Dory, raw	0	0	Tr	Tr	0	0	Tr	0.1	0	0	Tr	0.4
85	Lemon sole, steamed	0	0	Tr	Tr	Tr	0	0.2	0	0	0	Tr	0.1
92	Monkfish, raw	0	0	Tr	Tr	Tr	0	Tr	Tr	0	0	Tr	0.1
102	Plaice, raw[a]	0	0	Tr	Tr	Tr	Tr	0.2	0	0	0	Tr	0.1
103	grilled[a]	0	0	Tr	Tr	Tr	Tr	0.2	0	0	0	Tr	0.1
105	frozen, raw[a]	0	0	Tr	Tr	Tr	0	0	0	0	0	Tr	0.1
106	grilled[a]	0	0	Tr	Tr	Tr	Tr	0	0	0	0	0.1	0.1
108	-, steamed[a]	0	0	Tr	Tr	Tr	Tr	0	0	0	0	Tr	0.1
133	Rock Salmon/Dogfish, raw	0	0	0.1	0.1	0.1	0.1	0.8	0	0	0	0.2	1.4
142	Shark, raw	0	0	Tr	0	0	0	0.1	Tr	0	Tr	0.1	0.2
143	Skate, raw	0	0	Tr	Tr	Tr	Tr	Tr	0	0	0	Tr	0.1
159	Whiting, raw	0	0	Tr	Tr	Tr	0	0.1	0	0	0	Tr	0.1
160	steamed	0	0	Tr	Tr	Tr	0	0.1	0	0	0	Tr	0.1
Fatty Fish													
174	Eel, jellied	0	0	0.2	0.1	0	0	0.1	0.2	0	Tr	0.1	0.2
175	Herring, raw	0	0	0.3	0.2	0.3	0	Tr	0.8	0	0	0.1	1.0
176	grilled	0	0	0.2	0.2	0.3	0	Tr	0.7	0	0	0.1	0.8
187	Kipper, raw	0.1	0	0.2	0.3	0.5	Tr	0.1	1.1	0.1	0	0.1	1.3
188	grilled	0.1	0	0.2	0.3	0.6	Tr	0.1	1.3	0.1	0	0.1	1.5
191	Mackerel, raw	0	0	0.3	0.2	0.6	0.1	0.1	0.7	0	0	0.1	1.1
194	grilled	0	0	0.3	0.2	0.7	0.1	0.1	0.8	0	0	0.1	1.2

[a] Plaice contains 0.2g 22:2 per 100g

Fish and fish products

Fatty Acids per 100g food

No. 16-	Food	Saturated						Monounsaturated					
		12:0	14:0	15:0	16:0	17:0	18:0	14:1	16:1	17:1	18:1	20:1	22:1
201	**Pilchards**, canned in tomato sauce	0	0.3	0.1	0.9	0.1	0.4	Tr	0.5	0.1	1.1	0.3	0.2
202	**Salmon**, raw[a]	0	0.3	Tr	1.2	0	0.3	0	0.7	0	2.4	0.7	0.6
208	**Salmon, pink**, canned in brine, flesh only, *drained*	0	0.3	Tr	0.9	Tr	0.2	Tr	0.3	Tr	1.0	0.6	0.6
209	-, flesh and bones, *drained*	0	0.3	Tr	0.9	Tr	0.2	Tr	0.3	Tr	1.0	0.6	0.6
210	**Salmon, red**, canned in brine, flesh only, *drained*	0	0.4	Tr	1.2	Tr	0.2	Tr	0.4	Tr	1.4	1.0	0.9
211	-, flesh and bones, *drained*	0	0.3	Tr	0.8	Tr	0.2	Tr	0.3	Tr	1.0	0.8	0.7
212	**Sardines**, raw	0	0.7	Tr	1.5	0.1	0.3	Tr	0.9	Tr	0.7	0.3	0.6
213	*grilled*	0	0.8	0	1.7	0	0.4	Tr	1.0	Tr	0.8	0.4	0.7
216	canned in oil, *drained*	Tr	0.4	Tr	1.8	0.1	0.5	Tr	0.6	0.1	4.0	0.1	0.1
217	canned in tomato sauce	Tr	0.6	0.1	1.6	0.1	0.4	Tr	0.7	0.1	1.8	0.1	0.1
218	**Sprats**, raw	Tr	0.5	0.1	1.2	Tr	0.3	Tr	0.7	0.1	1.9	0.6	1.3
221	**Swordfish**, raw	0	0.2	Tr	0.5	0.3	0	0	0.3	0	1.2	0.1	0.1
225	**Trout, rainbow**, raw	Tr	0.2	Tr	0.7	Tr	0.2	Tr	0.3	Tr	1.0	0.3	0.3
226	*grilled*	Tr	0.2	Tr	0.7	Tr	0.2	Tr	0.3	Tr	1.1	0.3	0.3
228	**Tuna**, raw	0	0.1	Tr	0.7	Tr	0.4	Tr	0.2	Tr	0.8	0.1	0.1
229	canned in brine, *drained*	0	Tr	Tr	0.1	Tr	0.1	Tr	Tr	Tr	0.1	Tr	0
230	canned in oil, *drained*	Tr	Tr	Tr	0.8	Tr	0.6	0	Tr	Tr	2.2	Tr	0

[a] Salmon contains 0.1g 22:0 per 100g

Fish and fish products

No. Food 16-	Polyunsaturated											
	16:4	16:UNID	18:2	18:3	18:4	20:3	20:4	20:5	21:5	22:4	22:5	22:6
201 **Pilchards**, canned in tomato sauce	0.1	0	0.1	0.1	0.2	0	0.2	1.2	0.1	Tr	0.2	1.2
202 **Salmon**, raw	0	0	0.3	0.2	0.1	0.1	0.2	0.5	0	0	0.4	1.3
208 **Salmon, pink**, canned in brine, flesh only, *drained*	0	0	0.1	0.1	0.2	Tr	0.1	0.5	0	Tr	0.1	0.8
209 -, flesh and bones, *drained*	0	0	0.1	0.1	0.2	Tr	0.1	0.5	0	Tr	0.1	0.8
210 **Salmon, red**, canned in brine, flesh only, *drained*	0	0	0.1	0.1	0.2	Tr	0.1	0.7	0	Tr	0.2	0.9
211 -, flesh and bones, *drained*	0	0	0.1	0.1	0.2	Tr	0.1	0.5	0	Tr	0.1	0.7
212 **Sardines**, raw[a]	0.1	0	0.1	0.1	0.2	Tr	0.1	0.9	0	0	0.1	1.1
213 *grilled* [a]	0.1	0	0.1	0.1	0.2	0.1	0.1	1.0	Tr	0	0.1	1.2
216 canned in oil, *drained*	0.1	0	2.6	0.4	0.2	0	0.1	0.9	0	0	0	0.8
217 canned in tomato sauce	0.1	0	1.0	0.2	0.2	0	0.1	0.9	Tr	0	0.1	0.7
218 **Sprats**, raw	0.1	0	0.1	0.1	0.2	0	0.1	0	0.9	Tr	0.1	1.3
221 **Swordfish**, raw	0	0	0.1	Tr	Tr	0	0.1	0.1	0	Tr	0.2	0.5
225 **Trout, rainbow**, raw	Tr	0	0.3	0.1	0.1	0	0.1	0.2	Tr	0	0.1	0.8
226 *grilled*	Tr	0	0.4	0.1	0.1	0	0.1	0.2	Tr	0	0.1	0.7
228 **Tuna**, raw	0	0	0.1	0	Tr	0	0.1	0.3	0	Tr	0.1	1.1
229 canned in brine, *drained*	0	0	Tr	Tr	Tr	Tr	Tr	Tr	0	0	Tr	0.1
230 canned in oil, *drained*	0	0	3.4	0.9	0	0	Tr	0.1	0	0	Tr	0.3

a Sardines contain 0.1g 16:2 per 100g

Fish and fish products

Fatty Acids per 100g food

No.	Food	Saturated						Monounsaturated					
16-		12:0	14:0	15:0	16:0	17:0	18:0	14:1	16:1	17:1	18:1	20:1	22:1
Crustacea													
232	**Crab**, *boiled*	Tr	0.1	Tr	0.4	Tr	0.2	Tr	0.4	Tr	0.9	0.2	0.1
236	**Lobster**, *boiled*	0	Tr	0	0.1	0	Tr	0	0.1	0	0.2	Tr	Tr
239	**Prawns**, *boiled*	Tr	Tr	Tr	0.1	Tr	Tr	Tr	Tr	Tr	0.1	Tr	Tr
241	frozen, raw	Tr	Tr	Tr	0.1	Tr	Tr	Tr	Tr	Tr	0.1	Tr	Tr
245	**Shrimps**, *boiled*	0	Tr	Tr	0.3	Tr	Tr	0	0.1	Tr	0.3	Tr	Tr
247	canned in brine, *drained*	0	Tr	Tr	0.1	Tr	Tr	0	0	Tr	0.2	Tr	Tr
Molluscs													
251	**Clams**, canned in brine, *drained*	0	Tr	0	0.1	0	Tr	0	0.1	0	0.1	0	0
252	**Cockles**, *boiled*	Tr	Tr	Tr	0.1	Tr	Tr	0	Tr	0	Tr	Tr	0
253	bottled in vinegar, *drained*	Tr	Tr	Tr	0.1	Tr	0.1	0	Tr	0	Tr	Tr	0
254	**Cuttlefish**, raw	0	Tr	0	0.1	0	0.1	0	Tr	0	Tr	0.1	Tr
255	**Mussels**, raw	Tr	0.1	Tr	0.2	Tr	Tr	0	0.1	Tr	0.1	0.1	0
256	*boiled*	Tr	0.1	Tr	0.3	Tr	Tr	0	0.2	Tr	0.1	0.1	0
258	canned and bottled, *drained*	Tr	0.1	Tr	0.3	Tr	Tr	0	0.2	Tr	0.1	0.1	0
262	**Scallops**, *steamed*	0.1	Tr	Tr	0.2	Tr	Tr	0	Tr	0	0	Tr	Tr
263	**Squid**, raw	0	Tr	0	0.3	Tr	Tr	Tr	Tr	Tr	0.1	0.1	Tr
264	frozen, raw	0	Tr	Tr	0.2	Tr	Tr	Tr	Tr	Tr	0.1	0.1	Tr
268	**Whelks**, *boiled*	0	Tr	Tr	0.1	Tr	0.1	Tr	Tr	Tr	0.1	0.1	Tr
270	**Winkles**, *boiled*	Tr	0.1	Tr	0.1	Tr	Tr	Tr	Tr	Tr	0.1	0.1	Tr

Fish and fish products

Fatty Acids per 100g food

No. 16-	Food	Polyunsaturated 16:4	16:UNID	18:2	18:3	18:4	20:3	20:4	20:5	21:5	22:4	22:5	22:6
	Crustacea												
232	**Crab**, *boiled* [a]	0	0.1	Tr	Tr	0.1	Tr	0.1	0.5	Tr	Tr	0.1	0.5
236	**Lobster**, *boiled*	0	0	Tr	0.1	0	0	Tr	0.2	0	0	Tr	0.1
239	**Prawns**, *boiled*	0	0	Tr	Tr	Tr	0	Tr	0.1	0	0	Tr	0.1
241	frozen, raw	0	0	Tr	Tr	Tr	0	Tr	0.1	0	0	Tr	Tr
245	**Shrimps**, *boiled*	0	0	Tr	Tr	Tr	0	Tr	0.4	0	Tr	Tr	0.3
247	canned in brine, *drained*	0	0	Tr	Tr	Tr	0	Tr	0.2	0	Tr	Tr	0.1
	Molluscs												
251	**Clams**, canned in brine, *drained*	0	0	Tr	Tr	Tr	0	Tr	0.1	0	0	Tr	Tr
252	**Cockles**, *boiled*	0	Tr	Tr	0	Tr	0	Tr	Tr	0	Tr	Tr	Tr
253	bottled in vinegar, *drained*	0	Tr	Tr	0	Tr	0	Tr	0.1	0	Tr	Tr	Tr
254	**Cuttlefish**, raw	0	0	Tr	Tr	Tr	0	Tr	0.1	0	0	Tr	0.1
255	**Mussels**, raw	Tr	0.1	Tr	Tr	Tr	0	Tr	0.3	Tr	0	Tr	0.1
256	*boiled*	Tr	0.1	Tr	Tr	Tr	0	0.1	0.4	Tr	0	Tr	0.2
258	canned and bottled, *drained*	Tr	0.1	Tr	Tr	Tr	0	Tr	0.3	Tr	0	0	0.1
262	**Scallops**, *steamed*	0	0	Tr	Tr	Tr	0	Tr	0.1	0	0	Tr	0.1
263	**Squid**, raw	0	0	0.1	Tr	Tr	Tr	Tr	0.1	Tr	0	Tr	0.3
264	frozen, raw	0	0	0.1	Tr	Tr	Tr	Tr	0.1	Tr	0	Tr	0.3
268	**Whelks**, *boiled*	Tr	Tr	Tr	Tr	Tr	0	Tr	0	Tr	Tr	0.1	0.1
270	**Winkles**, *boiled*	Tr	Tr	Tr	0.1	Tr	0	0.1	0.1	Tr	Tr	Tr	Tr

[a] Crab contains 0.1g 22:UNID per 100g

109

Fish and fish products

Fatty Acids per 100g food

No. 16-	Food	Saturated						Monounsaturated					
		12:0	14:0	15:0	16:0	17:0	18:0	14:1	16:1	17:1	18:1	20:1	22:1
	Fish products and dishes												
272	**Caviare**, bottled in brine, *drained*	0	0.1	0	0.6	0	0.1	0	0.1	0	0.9	0.1	Tr
280	**Fish cakes**, frozen	Tr	Tr	Tr	0.4	Tr	0.1	0	Tr	Tr	1.2	Tr	Tr
287	**Fish fingers**, cod, frozen	Tr	0.1	0	2.1	0	0.3	0	Tr	0	3.0	0	0
293	**Fish paste**	0	0	0	0	0	0	0	0	0	0	0	0
299	**Roe**, cod, hard, raw	0	Tr	Tr	0.3	Tr	Tr	Tr	0.1	Tr	0.3	Tr	Tr
302	herring, soft, raw	0	Tr	Tr	0.4	Tr	0.1	0	Tr	0	0.4	0.1	Tr
307	**Taramasalata**	0	0.1	0	2.9	0	1.0	0	0.2	0	28.1	0.8	0.3

Fish and fish products

Fatty Acids per 100g food

No. Food	Polyunsaturated													
16-	16:4	16:UNID	18:2	18:3	18:4	20:3	20:4	20:5	21:5	22:4	22:5	22:6
Fish products and dishes												
272 **Caviare**, bottled in brine, *drained*	0	0	0.2	Tr	0	0	Tr	0.5	0	0	0.1	0.8
280 **Fish cakes**, frozen	0	0	0.5	0.2	Tr	Tr	Tr	Tr	0	0	Tr	0.1
287 **Fish fingers**, cod, frozen	0	0	1.8	0.1	0	0	0	0.1	0	0	0	0.1
293 **Fish paste**	0	0	0	0	0	0	0	0	0	0	0	0
299 **Roe**, cod, hard, raw	Tr	0	Tr	Tr	Tr	0	Tr	0.2	Tr	Tr	Tr	0.3
302 herring, soft, raw	0	0	Tr	Tr	Tr	0	Tr	0.2	0	0	Tr	0.4
307 **Taramasalata**	0	0	11.3	5.2	0	0	0	0.2	0	0	0.1	Tr

INDIVIDUAL SUGARS

Fish itself contains negligible amounts of sugars, although some glucose may be present in shellfish from the breakdown of glycogen. Where measurable quantities of individual sugars are present from batter and other coatings, sauces and recipe ingredients, the amounts are shown below. Values are given as their monosaccharide equivalents.

Individual sugars, g per 100g food

No. Food 16-	Glucose	Fructose	Sucrose	Maltose	Lactose
White fish					
25 **Cod**, coated in batter, frozen, *baked*	Tr	Tr	0.1	Tr	Tr
26 -, *weighed with bones*	Tr	Tr	0.1	Tr	Tr
27 coated in crumbs, frozen, *fried in blended oil*	0.1	Tr	Tr	Tr	0.1
47 **Haddock**, *poached*	0	0	0	0	1.1
48 -, *weighed with bones*	0	0	0	0	0.9
66 smoked, *poached*	0	0	0	0	1.1
67 -, *weighed with bones and skin*	0	0	0	0	1.0
76 **Halibut**, *poached*	0	0	0	0	1.1
77 -, *weighed with bones and skin*	0	0	0	0	1.0
88 **Lemon sole**, goujons, *fried in blended oil*	0	0	Tr	Tr	0.4
89 -, *fried in lard*	0	0	Tr	Tr	0.4
90 -, *fried in sunflower oil*	0	0	Tr	Tr	0.4
120 **Plaice**, goujons, *fied in blended oil*	0	0	Tr	Tr	0.4
121 -, *fried in lard*	0	0	Tr	Tr	0.4
122 -, *fried in sunflower oil*	0	0	Tr	Tr	0.4
Fatty fish					
182 **Herring**, canned in tomato sauce	0.8	1.0	1.1	Tr	Tr
198 **Mackerel**, canned in tomato sauce	0.5	0.6	0.3	Tr	Tr
201 **Pilchards**, canned in tomato sauce	0.4	0.5	Tr	0	0
217 **Sardines**, canned in tomato sauce	0.6	0.8	Tr	0	0

No. Food 16-	Individual sugars				
	Glucose	Fructose	Sucrose	Maltose	Lactose

Crustacea

No. Food	Glucose	Fructose	Sucrose	Maltose	Lactose
265 **Squid**, in batter, *fried in blended oil*	0	0	0.1	Tr	2.0
266 -, *fried in sunflower oil*	0	0	0.1	Tr	2.0

Fish products and dishes

No. Food	Glucose	Fructose	Sucrose	Maltose	Lactose
274 **Curry**, fish, Bangladeshi	0.4	0.3	0.4	0	0
275 fish and vegetable, Bangladeshi	0.4	0.4	0.3	0	0
276 haddock, Bengali	1.2	1.0	0.8	0	0
277 herring, Bengali	1.2	1.0	0.8	0	0
278 prawn and mushroom	0.6	0.6	0.5	0	0
285 **Fish cakes**, cod, homemade	0.1	Tr	0.1	Tr	0.1
286 salmon, homemade	0.1	Tr	0.1	Tr	0.1
293 **Fish paste**	Tr	Tr	0.5	Tr	0
294 **Fish pie**	0.1	0.1	0.2	Tr	1.2
295 **Fisherman's pie**, retail	0.1	Tr	0.1	Tr	1.4
296 **Kedgeree**	0	0	0	0	0.1
297 **Mackerel pâté**, smoked	0.1	Tr	Tr	0	0.6
298 **Pilau**, prawn	0.3	0.2	0.2	0	0
304 **Salmon en croûte**, retail	0.1	Tr	0.1	0.5	0.2
308 **Tuna pâté**	0.1	0.1	0.2	0	0

RETINOL FRACTIONS

White fish and its products contain little vitamin A while most of the fatty fish are nutritionally significant sources. They may contain 13-cis-retinol in addition to the more common all-trans form. Certain fatty fish contain measurable amounts of retinaldehyde, which is also the main retinol source in cod roe.

Where the amount of retinol fractions in a fish or fish product is known, the values are shown below. Total retinol is also shown, as the sum of all-trans-retinol, 90% of the retinaldehyde, 75% of the 13-cis-retinol and 40% of the dehydroretinol, (Sivell et al., 1984[a]).

Retinol fractions, μg per 100g food

No. Food 16-	All-trans-retinol	13-cis-retinol	Retin-aldehyde	Dehydro-retinol	Total Retinol
White fish					
12 **Cod**, raw	2	0	0	0	2
19 frozen, raw	2	0	0	0	2
31 **Coley**, raw	4	0	0	0	4
34 frozen, raw	4	0	0	0	4
128 **Red snapper**, *fried in blended oil, weighed with bones and skin*	2	0	0	0	2
129 *-, fried in sunflower oil*	4	0	0	0	4
130 *-, weighed with bones and skin*	2	0	0	0	2
131 **Redfish**, raw	8	0	0	0	8
Fatty fish					
168 **Anchovies**, canned in oil, *drained*	41	21	0	0	57
174 **Eel**, jellied	110	0	0	0	110
175 **Herring**, raw	33	14	0	0	44
176 *grilled*	25	9	0	0	34
177 *-, weighed with bones and skin*	17	6	0	0	23
187 **Kipper**, raw	31	2	0	0	32
188 *grilled*	36	2	0	0	38
189 *-, weighed with bones*	23	1	0	0	24

[a] Sivell, L. M., Bull, N.L., Buss, D.H., Wiggins, R.A., Scuffam, D., Jackson, P.A. (1984). Vitamin A activity in foods of animal origin. *J. Sci. Fd. Agric.* **35**, 931-939

No. Food 16-	Retinol fractions				Total Retinol
	All *trans*-retinol	13-*cis*-retinol	Retin-aldehyde	Dehydro-retinol	
191 **Mackerel**, raw	37	10	0	0	45
192 *fried in blended oil*	39	6	0	0	43
193 *-, weighed with bones and skin*	28	4	0	0	31
194 *grilled*	40	11	0	0	48
195 *-, weighed with bones and skin*	37	10	0	0	44
196 smoked	25	8	0	0	31
197 canned in brine, *drained*	27	13	0	0	37
198 canned in tomato sauce	23	11	0	0	31
201 **Pilchards**, canned in tomato sauce	5	2	0	0	7
202 **Salmon**, raw	7	1	0	12	13
203 *grilled*	8	1	0	14	16
204 *-, weighed with bones and skin*	7	1	0	12	13
205 *steamed*	8	1	0	13	14
206 *-, weighed with bones and skin*	6	1	0	10	11
208 **Salmon, pink**, canned in brine, flesh only, *drained*	22	9	0	7	31
209 *-, flesh and bones, drained*	22	9	0	7	31
210 **Salmon, red**, canned in brine, flesh only, *drained*	38	14	0	10	52
211 *-, flesh and bones, drained*	22	9	0	7	31
217 **Sardines**, canned in tomato sauce	8	2	0	0	9
218 **Sprats**, raw	60	0	0	0	60
225 **Trout, rainbow**, raw	28	1	0	51	49
226 *grilled*	10	1	0	44	29
227 *-, weighed with bones and skin*	7	1	0	32	21
Crustacea					
248 **Shrimps**, frozen	1	0	0	1	2
Molluscs					
252 **Cockles**, *boiled*	8	0	36	0	40
253 bottled in vinegar, *drained*	10	0	45	0	51
Fish products and dishes					
297 **Mackerel pâté**, smoked	145	27	0	0	165
299 **Roe**, cod, hard, raw	19	1	66	0	79

VITAMIN E FRACTIONS

The vitamin E activity of foods can be derived from a number of different tocopherols and tocotrienols. In fish the major contributor is α-tocopherol but measurable amounts of other tocopherols may also contribute to vitamin E activity.

Where vitamin E is present, and the amount of each tocopherol was known, the values are shown below. For these fish and fish products, the total vitamin E activity is also shown as α-tocopherol equivalents, which has been taken as the sum of the α-tocopherol, 40% of the β-tocopherol, 10% of the γ-tocopherol and 1% of the δ-tocopherol (McClaughlin and Weihrauch 1979[a]).

No. 16-	Food	α-tocopherol	β-tocopherol	γ-tocopherol	δ-tocopherol	Vitamin E equiv
	White fish					
7	**Catfish**, raw	2.10	0	0	0	2.10
8	*steamed*	2.80	0	0	0	2.80
9	*-, weighed with bones*	2.38	0	0	0	2.38
12	**Cod**, raw	0.44	0	0	0	0.44
13	*baked*	0.59	0	0	0	0.59
14	*-, weighed with bones and skin*	0.50	0	0	0	0.50
15	*poached*	0.61	0	0	0	0.61
16	*-, weighed with bones and skin*	0.53	0	0	0	0.53
17	*steamed*	0.48	0	0	0	0.48
18	*-, weighed with bones and skin*	0.39	0	0	0	0.39
19	*frozen, raw*	(0.44)	0	0	0	(0.44)
20	*-, grilled*	(1.00)	0	0	0	(1.00)
31	**Coley**, raw	0.36	0	0	0	0.36
32	*steamed*	0.46	0	0	0	0.46
33	*-, weighed with bones and skin*	0.39	0	0	0	0.39
34	*frozen, raw*	0.36	0	0	0	0.36
38	**Dab**, raw	0.40	0	0	0	0.40
44	**Haddock**, raw	0.39	0	0	0	0.39
45	*grilled*	0.50	0	0	0	0.50
46	*-, weighed with bones*	0.47	0	0	0	0.47

Vitamin E fractions, mg per 100g food

[a] McClaughlin, T.J. and Weihrauch, J.L. (1979). Vitamin E content of foods. *J Am. Diet. Assoc.* **75**, 647-665

No. 16-	Food	α-tocopherol	β-tocopherol	γ-tocopherol	δ-tocopherol	Vitamin E equiv
47	**Haddock,** *poached*	0.35	0	0	0	0.44
48	-, *weighed with bones*	0.30	0	0	0	0.37
49	*steamed*	0.41	0	0	0	0.41
50	-, *weighed with bones and skin*	0.34	0	0	0	0.34
51	frozen, raw	0.35	0	0	0	0.35
73	**Halibut,** raw	0.85	0	0	0	0.85
74	*grilled*	1.00	0	0	0	1.00
75	-, *weighed with bones and skin*	0.78	0	0	0	0.78
76	*poached*	0.95	0	0	0	1.04
77	-, *weighed with bones and skin*	0.87	0	0	0	0.95
91	**Ling,** raw	0.30	0	0	0	0.30
98	**Mullet, Red,** raw	0.51	Tr	Tr	0	0.51
99	*grilled*	0.59	Tr	Tr	0	0.59
100	-, *weighed with bones and skin*	0.28	Tr	Tr	0	0.28
127	**Red snapper,** *fried in blended oil*	1.30	0.12	0.04	0	1.35
128	-, *weighed with bones and skin*	0.66	0.06	0.02	0	0.69
131	**Redfish,** raw	1.25	0	0	0	1.25
	Fatty fish					
172	**Carp,** raw	0.63	0	0	0	0.63
174	**Eel,** jellied	2.60	0	0	0	2.60
175	**Herring,** raw	0.76	Tr	Tr	0	0.76
176	*grilled*	0.64	Tr	Tr	0	0.64
177	-, *weighed with bones and skin*	0.44	Tr	Tr	0	0.44
182	canned in tomato sauce	3.27	0.05	2.73	0.47	3.57
187	**Kipper,** raw	0.32	0	0	0	0.32
188	*grilled*	0.37	0	0	0	0.37
189	-, *weighed with bones*	0.23	0	0	0	0.23
191	**Mackerel,** raw	0.43	0	0	0	0.43
194	*grilled*	0.46	0	0	0	0.46
195	-, *weighed with bones and skin*	0.42	0	0	0	0.42
196	smoked	0.25	0	0	0	0.25
198	canned in tomato sauce	1.92	0.05	Tr	Tr	1.94
201	**Pilchards,** canned in tomato sauce	2.50	0.13	0.04	0	2.56

No. Food 16-	α- tocopherol	β– tocopherol	γ– tocopherol	δ– tocopherol	Vitamin E equiv
202 **Salmon**, raw	1.90	0	0.15	0	1.91
203 *grilled*	2.30	0	0.18	0	2.29
204 -, *weighed with bones and skin*	1.89	0	0.15	0	1.88
205 *steamed*	2.05	0	0.16	0	2.07
206 -, *weighed with bones and skin*	1.58	0	0.12	0	1.59
208 **Salmon, pink**, canned in brine, flesh only, *drained*	1.52	0	0.03	0	1.52
209 -, flesh and bones, *drained*	1.52	0	0.03	0	1.52
210 **Salmon, red**, canned in brine, flesh only, *drained*	2.08	0	0.04	0	2.08
211 -, flesh and bones, *drained*	1.52	0	0.03	0	1.52
212 **Sardines**, raw	0.28	0.03	Tr	Tr	0.29
213 *grilled*	0.31	0.03	0	0	0.32
214 -, *weighed with bones*	0.19	0.02	0	0	0.20
216 canned in oil, *drained*	0.26	0	0.44	1.10	0.31
217 canned in tomato sauce	2.90	0.27	0.65	0.32	3.08
218 **Sprats**, raw	0.51	0	0	0	0.51
225 **Trout, rainbow**, raw	0.71	0	0.05	0	0.71
226 *grilled*	1.00	0	0.06	0	1.01
227 -, *weighed with bones and skin*	0.73	0	0.04	0	0.74
229 **Tuna**, canned in brine, *drained*	0.55	0	0	0	0.55
230 canned in oil, *drained*	1.30	0.19	5.40	2.20	1.94
Crustacea					
255 **Mussels**, raw	0.74	0	0	0	0.74
256 *boiled*	1.05	0	0	0	1.05
257 -, *weighed with shells*	0.28	0	0	0	0.28
260 **Oysters**, raw	0.85	0	0	0	0.85
261 -, *weighed with shells*	0.12	0	0	0	0.12
263 **Squid**, raw	1.20	0	0	0	1.20
268 **Whelks**, *boiled*	0.80	0	0	0	0.80
269 -, *weighed with shells*	0.27	0	0	0	0.27
270 **Winkles**, *boiled*	3.90	0	0	0	3.90
271 -, *weighed with shells*	0.74	0	0	0	0.74

No. Food 16-	α-tocopherol	β-tocopherol	γ-tocopherol	δ-tocopherol	Vitamin E equiv
Fish products and dishes					
293 **Fish paste**	0.87	0	0	0	0.87
297 **Mackerel pâté**, smoked	0.93	0.27	0.75	0.23	1.11
299 **Roe**, cod, hard, raw	6.20	0	0.01	0	6.20
302 herring, soft, raw	2.80	0	0.01	0	2.80

RECIPES

Unless specified the recipes use whole pasteurised milk, fresh cream, Cheddar cheese and plain white flour.

An average egg has been assumed to weigh 50g. A level teaspoon refers to a standard 5 ml spoon and has been taken to hold 5g salt and 3g spices.

The type of fat used in the recipes has been specified. The vegetable oil was a retail blended vegetable oil. Margarine was an average of hard, soft and polyunsaturated types. The butter was salted.

For fried dishes the fat absorbed during frying has been included at the end of the ingredients list with the quantity absorbed shown in brackets.

274 Curry, fish, Bangladeshi

450g boal, raw	2 tsp chilli powder
450g rohu, raw	2 tsp coriander powder
225g onions, chopped	½ tsp cumin powder
75g vegetable oil	1½ tsp turmeric
2 tsp salt	300g water

Cut the fish into 1 inch slices and sprinkle with some of the chilli, turmeric and coriander. Add 2 tbsps of water and mix. Heat half the oil and fry the fish for 6 to 8 minutes then remove from pan and set aside. Fry the onions in the remaining oil until brown, add remaining spices and the remaining water and cook for 6 minutes. Add fish and salt and cook for 4 to 5 minutes. Add water, cover and cook for 10 minutes.

Weight loss: 21%

275 Curry, fish and vegetable, Bangladeshi

225g boal, raw	1½ tsp chilli powder
225g hilsa, raw	1 tsp coriander powder
225g mackerel, raw	1 tsp cumin
225g rohu, raw	1 tsp turmeric
200g onions, chopped	2 tsp salt
225g courgettes, sliced	600g water
225g green beans, sliced	20g coriander leaves, chopped
70g vegetable oil	

Cut the fish into 1 inch slices. Fry the onions in the oil until brown. Add chilli, turmeric, coriander and cumin with 100ml water and cook for 5 to 6 minutes. Add fish and cook gently for 5 minutes. Add the vegetables and salt, cook with stirring for 5 to 6 minutes. Add remaining water, bring to the boil and simmer for 10 to 15 minutes until vegetables are soft. Sprinkle with coriander leaves.

Weight loss: 15%

276 Curry, haddock, Bengali

450g haddock, raw	5g ginger root, chopped
300g onions, chopped	5g green chillies, chopped
200g tomatoes	4g mixed spice
200g butter	2g garlic, crushed
7g salt	100g water

Fry onion in butter until brown. Add ginger, garlic, chillies, spices and salt together with a little water. Add tomatoes and remaining water, simmer for a few minutes. Coat fish on both sides with cooked mixture. Bake at 200°C/mark 6 for 25 minutes.

Weight loss: 40%

277 Curry, herring, Bengali

450g herring, raw	5g ginger root, chopped
300g onions, chopped	5g green chillies, chopped
200g tomatoes	4g mixed spice
200g butter	2g garlic, crushed
7g salt	100g water

Fry onion in butter until brown. Add ginger, garlic, chillies, spices and salt together with a little water. Add tomatoes and remaining water, simmer for a few minutes. Coat fish on both sides with cooked mixture. Bake at 200°C/mark 6 for 25 minutes.

Weight loss: 40%

278 Curry, prawn and mushroom

110g prawns, raw	10g green chillies
225g mushrooms, sliced	10g tomato purée
90g onions, chopped	½ tsp chilli powder
60g vegetable oil	½ tsp cumin
10g garlic	½ tsp tumeric
10g ginger root	½ tsp salt

Liquidise garlic, ginger and chillies. Heat oil, add onion and fry until brown. Add tomato puree, salt, spices and liquidised ingredients and cook for a further minute. Add mushrooms and prawns and simmer gently for 8 minutes.

Weight loss: 18%

285 Fish cakes, cod, homemade

200g cod, cooked	25g flour
200g potatoes, mashed	½ tsp salt
2 eggs	
50g breadcrumbs	
vegetable oil absorbed on frying (80g)	

Mix together fish, one egg, potatoes and salt. Shape into 6 cakes. Dip in remaining egg and coat with flour and breadcrumbs. Shallow fry until golden brown.

Weight loss: 5%

286 Fish cakes, salmon, homemade

200g salmon, cooked   25g flour
200g potatoes, mashed   ½ tsp salt
2 eggs
50g breadcrumbs
vegetable oil absorbed on frying (80g)

Mix together fish, one egg, potatoes and salt. Shape into 6 cakes. Dip in remaining egg and coat with flour and breadcrumbs. Shallow fry until golden brown.

Weight loss: 5%

294 Fish pie

200g cod, cooked   15g flour
400g potatoes, mashed   15g margarine
150g milk   ½ tsp salt

Make a white sauce with milk, margarine, flour and salt. Mix sauce with flaked fish. Pipe a potato border round a dish, pour in the fish mixture. Brown in the oven at 200°C/mark 6 for 30 minutes.

Weight loss: 10%

296 Kedgeree

200g smoked haddock, cooked   25g margarine
100g rice, boiled   ½ tsp salt
2 eggs

Hard boil one egg. Melt the margarine and stir in the haddock, rice, salt and one beaten egg. Stir in chopped hard boiled egg and heat thoroughly.

Weight loss: 10%

298 Pilau, prawn

225g prawns, raw   25g ginger root, chopped
400g rice, long grain   10g garlic, crushed
180g onions, chopped   1 tsp cumin seeds
50g butter   1 tsp garam masala
1 tsp salt   1000g water

Soak the rice. Heat butter and add garlic, onions and ginger, fry until brown. Add drained rice and prawns and cook for 5 minutes. Add remaining ingredients, bring to the boil then reduce heat, cover pan and simmer for 15 minutes.

Weight loss: 23%

ALTERNATIVE AND TAXONOMIC NAMES

Foods are listed below in the same order as in the main tables.

The alternative names listed in the left-hand column below are those that were most frequently encountered during data collection and are included to help identify foods. It is important to recognise that in some cases such names may be used for more than one food and that all such usages may not appear in this list.

To see if a name is listed the food index should be consulted first. If the term is included as an alternative name, a cross reference entry indicates the food entry to which it refers.

Taxonomic names listed in the right-hand column refer as specifically as possible to the data used. Where two or more taxonomic names are listed, the data are representative of a mixture of these varieties.

The abbreviation 'sp' and 'spp' are used to indicate that one or more than one species of the specified Genus is included.

Alternative Names	Food Names	Taxonomic Names
Ormer	**Abalone**	*Haliotis gigantea*
	Anchovies	*Engraulis encrasicolus*
Air Ari Shede	**Ayr**	*Mystus seenghala*
Bachura Jhalli	**Bacha**	*Eutropiichthys vacha*
Common bass	**Bass, Sea**	*Dicentrachus labrax*
	Bele	*Glassogobius giuris*
	Bloater	*Clupea harengus*
Boalee Bombli	**Boal**	*Wallago attu*

Alternative Names	Food Names	Taxonomic Names
Bumalo Bummalow Sukki chhoti machhi	**Bombay duck**	*Harpodon nehereus*
Scup	**Bream, Sea**	*Stenotomus chrysops*
	Carp	*Cyprinus carpio*
Atlantic catfish Rockfish Wolf-fish	**Catfish**	*Anarhichas lupus*
Featherback	**Chital**	*Notopterus chitala*
Short-neck clam	**Clams**	*Venerupis semidecussata*
	Cockles	*Cardium edule*
	Cod	*Gadus morhua*
Coalfish Saithe	**Coley**	*Pollachius virens*
	Conger eel	*Conger conger*
	Crab	*Cancer pagurus*
Ocean sticks	**Crabsticks**	
	Crayfish	*Astacus sp* *Procrambus sp*
	Cuttlefish	*Sepia sp*
Common dab Dab sole	**Dab**	*Limanda limanda*
Sole	**Dover sole**	*Solea solea* *Solea vulgaris*

Alternative Names	Food Names	Taxonomic Names
	Eel	*Anguilla anguilla*
European flounder Fluke	**Flounder**	*Platichthys flesus*
	Flying fish	*Cypselurus spp* *Exocoetus spp* *Hirudichthyes spp*
	Haddock	*Melanogrammus aeglefinus*
	Hake	*Merluccius merluccius*
Atlantic halibut Butt	**Halibut**	*Hippoglossus hippoglossus*
	Herring	*Clupea harengus*
Pala Paliya Paluva Sboor	**Hilsa**	*Clupea ilisha*
	Hoki	*Macruronus novaezealandiae*
Crevalle Jack Trevally Jack	**Jackfish**	*Caranx spp*
Dory Peter-fish	**John Dory**	*Zeus faber*
Kala-beinse Kalvus	**Kalabasu**	*Labeo calbasu*
Barkur Karakatla Tambra	**Katla**	*Catla catla*
	Kipper	*Clupea harengus*

Alternative Names	Food Names	Taxonomic Names
Lemon dab Lemon fish Sweet fluke	**Lemon sole**	*Microstromus kitt*
Common ling	**Ling**	*Molva molva*
	Lobster	*Homarus vulgaris*
	Mackerel	*Scomber scombrus*
Anglerfish Goosefish	**Monkfish**	*Lophius piscatorius*
Striped mullet	**Mullet, Grey**	*Mugil cephalus*
Goatfish	**Mullet, Red**	*Mullus surmuletus*
Blue mussel	**Mussels**	*Mytilis edulis*
	Octopus	*Octopus vulgaris*
Sea perch	**Orange roughy**	*Hoplostethus atlanticus*
Common oyster Flat oyster	**Oysters**	*Ostrea edulis*
Pangsa	**Pangas**	*Pangasius pangasius*
	Parrot fish	*Scarus nuchipunctatus* *Scarus sparisoma*
	Pilchards	*Sardina pilchardus*
	Plaice	*Pleuronectes platessa*
Walleye pollack	**Pollack, Alaskan**	*Theragra chalcogramma*
Butterfish Chinese pomfret Rupchanda	**Pomfret**	*Formio niger* *Stromateus niger* *Stromateus sinensis*

Alternative Names	Food Names	Taxonomic Names
Chingree Esha Haunya Jheenga	**Prawns**	*Paleamon serratus*
Malabar	**Red snapper**	*Lutjanus sp*
Ocean perch Red perch Sea bream	**Redfish**	*Sebastes marinus*
Dogfish Huss Rock eel Spurdog Tope	**Rock salmon**	*Scyliorhinus caniculus* *Squalus acanthias*
Ruee Rui Tambada-massa	**Rohu**	*Labeo rohita*
	Salmon, pink	*Oncorhynchus gorbuscha*
Atlantic salmon	**Salmon**	*Salmo salar*
Sockeye salmon	**Salmon, red**	*Oncorhynchus nerka*
	Sardines	*Sardina pilchardus*
Common scallop	**Scallops**	*Pecten sp*
Dublin bay prawn Langoustine Norway Lobster	**Scampi**	*Nephrops norvegicus*
	Shark	*Lamna nasus*
	Shrimps	*Crangon crangon*
Ray Roker	**Skate**	*Raja spp*

Alternative Names	Food Names	Taxonomic Names
Brisling	**Sprat**	*Sprattus sprattus*
Calamari	**Squid**	*Loligo vulgaris*
	Swordfish	*Xiphias gladius*
Bolti Bulti Nile perch	**Tilapia**	*Tilapia nilotica*
	Trout, brown	*Salmo trutta*
	Trout, rainbow	*Salmo gairdneri*
	Tuna, canned	*Euthynnus pelamis*
Tunny	**Tuna**	*Thunnus spp*
Britt Butt	**Turbot**	*Psetta maxima* *Scopthalmus maximus*
Buckie	**Whelks**	*Buccinum indatum*
	Whitebait	Young of *Clupea harengus* Young of *Sprattus sprattus*
	Whiting	*Merlangius merlangus*
Buckie Periwinkle	**Winkles**	*Littorina littorea*

REFERENCES TO TABLES

1. Exler, J. (1987) *Composition of foods: finfish and shellfish products, raw, processed and prepared*, Agriculture Handbook No 8-15, US Department of Agriculture, Washington DC

2. Gopalan, C., Rama Sastri, B.V. and Balasubramanian, S.C. (1980) *Nutritive value of Indian foods*, National Institute of Nutrition, Indian Council of Medical Research, Hyderabad

3. Lewis, J. and English, R. (1990) *Composition of Foods*, Australia. Volume 3, Dairy Products, Eggs and Fish. Department of Community Services and Health, Canberra

4. Souci-Fachman-Kraut (1989) *Food composition and nutrition tables* 1989/90, 4th revised and completed edition. Wissenschaftliche Verlagsgesellschaft mbH, Stuttgart

5. Wu Leung, W.T., Butrum, R.R., Chang, F H., Narayana Rao, M. and Polacchi, W. (1972) *Food composition table for use in East Asia*, Food and Agriculture Organization and US Department of Health, Education and Welfare, Bethesda

A full list of the 244 references consulted is available on request

FOOD INDEX

Foods are indexed by their food number and **not** by page number.

Cross references in this index (e.g. Goosefish see **Monkfish**) give access to the individual food items through this index and to alternative and taxonomic names given on pages **123**.

Abalone, canned in brine, drained	250	Butt	see **Turbot**
Air	see **Ayr**	Butterfish	see **Pomfret**
Alaskan pollack	see **Pollack, Alaskan**		
Anchovies, canned in oil, drained	168	Calamari	see **Squid**
Anglerfish	see **Monkfish**	Carp, raw	172
Ari	see **Ayr**	Catfish	7–9
Atlantic catfish	see **Catfish**	Catfish, raw	7
Atlantic halibut	see **Halibut**	Catfish, steamed	8
Atlantic salmon	see **Salmon, raw**	Catfish, steamed, weighed with bones	9
Ayr, raw	1	Caviare, bottled in brine, drained	272
		Chinese pomfret	see **Pomfret**
Bacha, raw	169	Chinese salted fish, steamed	10
Bachura	see **Bacha**	Chingree	see **Prawns**
Barkur	see **Katla**	Chital, raw	11
Bass, Sea, raw	2	Clams, canned in brine, drained	251
Bele, raw	3	Coalfish	see **Coley**
Black pomfret	see **Pomfret, black**	Cockles	252–253
Bloater,	170–171	Cockles, boiled	252
Bloater, grilled	170	Cockles, bottled in vinegar, drained	253
Bloater, grilled, weighed with bones and skin	171	Cod	12–30
Blue mussel	see **Mussels**	Cod, baked	13
Boal, raw	4	Cod, baked, weighed with bones and skin	14
Boalee	see **Boal**	Cod, coated in batter, frozen, baked	25
Bolti	see **Tilapia**	Cod, coated in batter, frozen, baked,	
Bombay duck	5	weighed with bones	26
Bombli	see **Boal**	Cod, coated in crumbs, frozen, fried	27
Bream, Sea, raw	6	Cod, frozen, grilled	20
Brown trout	see **Trout, brown**	Cod, frozen, raw	19
Brisling	see **Sprat**	Cod, in batter, fried in blended oil	21
Britt	see **Turbot**	Cod, in batter, fried in dripping	22
Buckie	see **Whelks**	Cod, in batter, fried in retail blend oil	23
Buckie	see **Winkles**	Cod, in batter, fried in sunflower oil	24
Bulti	see **Tilapia**	Cod, in parsley sauce, frozen, boiled	30
Bumalo	see **Bombay duck**	Cod, poached	15
Bummalow	see **Bombay duck**	Cod, poached, weighed with bones and skin	16
Butt	see **Halibut**	Cod, raw	12